Understanding Evolution

E. Peter Volpe
Tulane University

Third Edition

Wm. C. Brown Company Publishers
Dubuque, Iowa

Library of Congress Catalog Card Number: 76–22978

ISBN 0–697–04546–3

Fifth Printing, 1979

Printed in the United States of America

To My
Ever-Probing Students
Who Taught Me
How to Teach

Contents

Preface

This new edition of *Understanding Evolution,* like the earlier ones, is addressed to the students for whom it is written. I have endeavored to present a simple, concise account of the scope and significance of evolution for the college student seeking a liberal education. I hope that it may dispel the vague and naive notions about organic evolution held by many beginning students. They view evolution as something that happened in the remote past. They know it has something to do with dinosaurs, rocks, and the proverbial "missing link." And they are familiar with such alluring clichés as "struggle for existence" and "survival of the fittest." But few pause to consider how the process of evolution actually works. In the belief that any concept can be best understood by knowing how it operates, I have focused on the mechanism of evolution, the causal aspect rather than the historical. Particular attention has been given to modern observations and experiments that illustrate and clarify the evolutionary process. Several examples of changes in human populations have been cited, so that the student may realize that evolution in mankind has not come to a standstill.

Evolution is a process of continual change. Organisms throughout life's history have not remained constant but have gradually and endlessly changed. Change is the rule of living things. The occurrence of organic evolution does not in itself reveal *how* evolution is brought about. That is, we may know of an event or phenomenon and accept it as true, even though we may not fully understand the forces that determine its existence. Scientists no longer debate the reality of evolution. However, differences of opinion have arisen in the *explanation* of evolution. One may challenge an interpretation, but to contest the interpretation is not to deny the existence of the event itself. A widespread fallacy is to discredit the truth of evolution by seizing on points of disagreement concerning the mechanism of evolution.

Charles Darwin was the first person to reach a meaningful understanding of the mechanism of evolution. The concept of natural selection that he masterly put forward in *The Origin of Species,* in 1859, remains the keystone of the evolutionary process. The Darwinian principle of natural selection is now firmly established as the main driving force of evolution. However, the Darwinian thesis has been enriched and refined by recent advances in systematics, ecology, cytology, and paleontology, and above all, in genetics. The modern extension of the great work of Gregor Mendel in heredity has profoundly influenced current thoughts about evolution. The mechanism of evolution is incomprehensible without familiarity with the fundamentals of genetics. Thus, the basic principles of genetics needed for an understanding of evolution are included in this text.

The opening chapter introduces the student to a specific population of organisms in which a novel trait has suddenly appeared among its members. An important consideration immediately imparted is that evolution is a property of populations. A population is not a mere assemblage of individuals; it is a breeding community of individuals. Moreover, it is the *population* that evolves in time, *not* the *individual.* An individual cannot evolve, for he survives for only one generation. An indivdual, however, can ensure continuity to a population by leaving offspring. Life is perpetuated only by the propagation of new individuals who differ in some degree from their parents. It will become apparent that the future success or failure of a population depends to a large extent on the numbers and kinds of offspring produced each generation. A deep appreciation of evolutionary changes in populations requires knowledge of mathematics. The general reader, however, should experience no difficulty with the mathematical considerations in this book, as the treatment is minimal and kept at an elementary level.

The text is simplified by numerous illustrations. The drawings, many of them original, are the accomplished work of my wife, Carolyn. The illustrations do not merely adorn the pages; all are important in supporting the writing. Selected references accompany each chapter in the hope that the student will want to extend his knowledge beyond the material presented.

The book could not have been prepared without the knowledge derived from my past teachers and students. The latter, perhaps unknowingly, have also been my teachers. If simplicity and clarity of presentation of the subject matter have been achieved in this book, it will be due in great measure to the long-suffering college freshmen in my introductory general biology course over the past fifteen years or so. I appreciate the helpful suggestions and advice from several colleagues, particularly Darrel L. Murray of the University of Virginia, Gerald Eck of the University of Washington (Seattle), and Donald Symons of the University of California (Santa Barbara). I wish also to acknowledge the able assistance of Helen Mequet, who cheerfully typed the manuscript in its various forms, and of Paula Chane Gebhardt, who worked tirelessly in preparing the Index and in reading proofs. Finally, I am indebted to the authors and publishers who have generously granted permission to use figures and tables from their books. Individual acknowledgments are made where the figures and charts appear in the text.

E. Peter Volpe

New Orleans, Louisiana

1 Meaning of Evolution

In the fall of 1958, the folks of a quiet, rural community in the southern part of the United States were startled and dismayed by the large numbers of *multilegged* bullfrogs in an 85-acre artificial lake on a cotton farm. Widespread newspaper publicity of this strange event attracted the attention of university scientists, curiosity seekers, and gourmets. The lake supports a large population of bullfrogs, estimated at several thousand. Although reports tended to be exaggerated, there were undoubtedly in excess of 350 multilegged deviants. As illustrated in figure 1.1, the extra legs were oddly positioned, but they were unmistakably copies of the two normal hind limbs. Incredibly the extra limbs were functional, but their movements were perceptibly not in harmony with the pair of normal legs. The bizarre multilegged frogs were clumsy and graceless.

All the multilegged frogs appeared to be of the same age, approximately two years old, and of the same generation. These atypic frogs were found only during the one season, and were not detected again in subsequent years. The multilegged frogs disappeared almost as dramatically as they had appeared.

Strange and exceptional events of this kind are of absorbing interest, and challenge us for an explanation. How do such oddities arise and what are the factors responsible for their ultimate disappearance in a natural population? In ancient times, bodily deformities evoked reverential awe and inspired some fanciful tales. Early man constructed a number of myths to explain odd events totally beyond his control or comprehension. One legend has it that when masses of skeletons are revived or reanimated, the bones of different animals often become confused. Another old idea is that grotesquely shaped frogs are throwbacks to some remote prehistoric ancestor. These accounts are, of course, novelistic and illusory. They do, however, reveal the uniquely imaginative capacity of the human mind. Nevertheless,

MULTILEGGED BULLFROGS

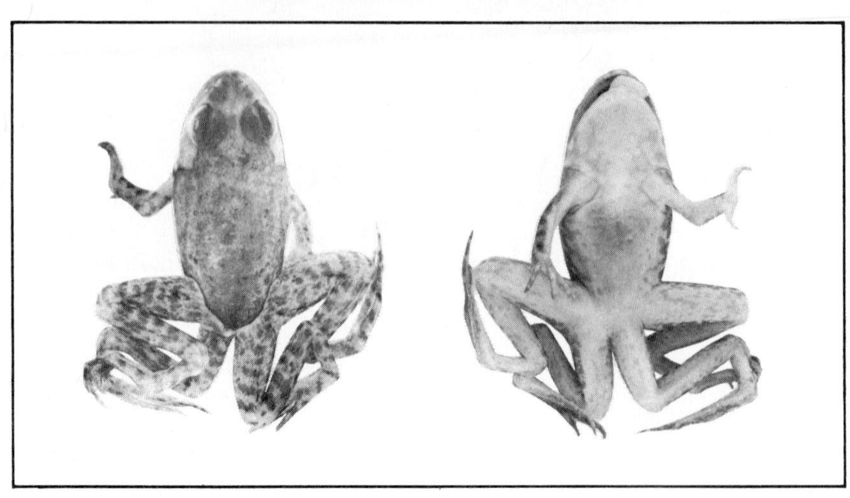

Figure 1.1 Two multilegged frogs, each viewed from the back (dorsal) and front (ventral) surface. These bizarre bullfrogs were discovered in October, 1958, in a lake near Tunica, Mississippi. Several hundred frogs with extra hind limbs were found at this locality. How does such an abnormality arise? Two reasonable interpretations are set forth in chapter 1.

we should seek a completely different cause-and-effect sequence, relying more on our faculty for logical analysis.

Environmental Modification

Inspection alone cannot reveal the underlying cause of the multilegged anomaly. We may thoroughly dissect the limbs and describe in detail the anatomy of each component, but no amount of dissection can tell us how the malformation arose. The deformity either was foreordained by heredity or originated from injury to the embryo at a vulnerable stage in its development. We shall direct our attention first to the latter possibility and its implications.

Some external factor in the environment may have adversely affected the pattern of development of the hind limb region. The cotton fields around the pond were periodically sprayed with pesticides to combat noxious insects. It is not inconceivable that the chemicals used were potent *teratogens* —that is, substances capable of causing marked distortions of normal body parts.

That chemical substances can have detrimental effects on the developing organism is well documented. For example, the geneticist Walter Landauer demonstrated in the 1950s that a wide variety of chemicals, such as boric acid, pilocarpine, and insulin (normally a beneficial hormone), can produce abnormalities of the legs and beaks when injected into chick embryos. This type of finding cannot be dismissed complacently as an instance of a laboratory demonstration without parallel in real life. Indeed, in 1961, medical researchers discovered with amazement amounting to incredulity that a purportedly harmless sleeping pill made of the drug thalidomide, when taken by a pregnant woman, particularly during the second month of pregnancy, could lead to a grotesque deformity in the newborn baby, a rare condition in humans called phocomelia—literally, "seal limbs." The arms are absent or reduced to tiny, flipperlike stumps (fig. 1.2).

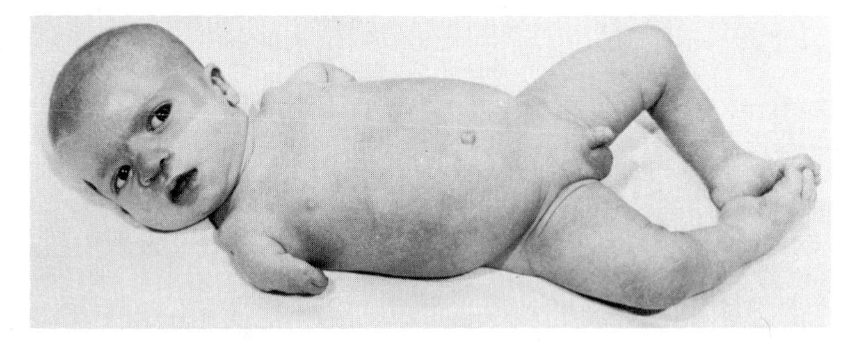

Figure 1.2 Armless deformity in the newborn infant, resulting from the action of thalidomide, a sedative taken by his mother in her second month of pregnancy. (Courtesy of Dr. W. Lenz.)

In different organisms—fish, frog, mice, and man—anomalies can be caused by such diverse agents as extremes of temperature, X rays, viruses, drugs, diet deficiencies, and lack of oxygen. More than a century ago, in 1832, the French biologist Etienne Geoffroy St. Hilaire sealed with varnish the air pores in the shell of a hen's egg, and observed that the embryo, deprived of its oxygen supply, became deformed. Experiments by modern investigators, like Theodore Ingalls of Harvard University, have confirmed St. Hilaire's crude but informative experiment. Deprivation of oxygen during the early divisions of fertilized eggs of the zebra fish can lead to the formation of abnormally small eyes or a single eye only. In 1941, the medical community was startled by the observation that pregnant women who had contracted German measles during the first three months of pregnancy often gave birth to blinded infants. The infants were afflicted with cataract, a condition in which the eye lens is opaque, obstructing the passage of light. There is no longer the slightest doubt that environmental factors may be causal agents of specific defects. A variation that arises as a direct response to some external change in the environment, and not by any change in the genetic makeup of the individual, is called an *environmental modification.*

If the multilegged anomaly in the bullfrog was environmentally induced, we may surmise (as depicted in figure 1.3) that the harmful chemical or other causative factor acted during the sensitive early embryonic stage. Moreover, if some external agent had brought about the abnormality in the bullfrog, this agent must have been effective only once, because the multilegged condition did not occur repeatedly over the years. It may be that the harmful environmental factor did not recur, in which event we would not expect a reappearance of the malformation.

Our suppositions could be put to a test by controlled breeding experiments, the importance of which cannot be overstated. Only through breeding tests can the basis of the variation be firmly established. If the anomaly constituted an environmental modification, then a cross of two multilegged bullfrogs would yield all normal progeny, as illustrated in figure 1.3. In the absence of any disturbing environmental factors, the offspring would develop normal hind limbs. This breeding experiment was not actually performed; none of the malformed frogs survived to sexual maturity. Nonetheless, we have brought into focus an important biological principle: *environmentally induced traits cannot be passed on to another generation.*

The elements that are transmitted to the next generation are two minute cells, the egg and the sperm. These two specialized cells are often referred to as *germ* cells because they are the beginnings, or germs, of new individuals. The germ cells represent the only connecting thread between successive generations. Accordingly, the mechanism of hereditary transmission must operate across this slender connecting bridge. The hereditary qualities of the offspring are established at the time the sperm unites with the egg. The basic hereditary determiners, the *genes,* occur in pairs in the fertilized egg. Each inherited characteristic is governed by at least one pair of genes, with one

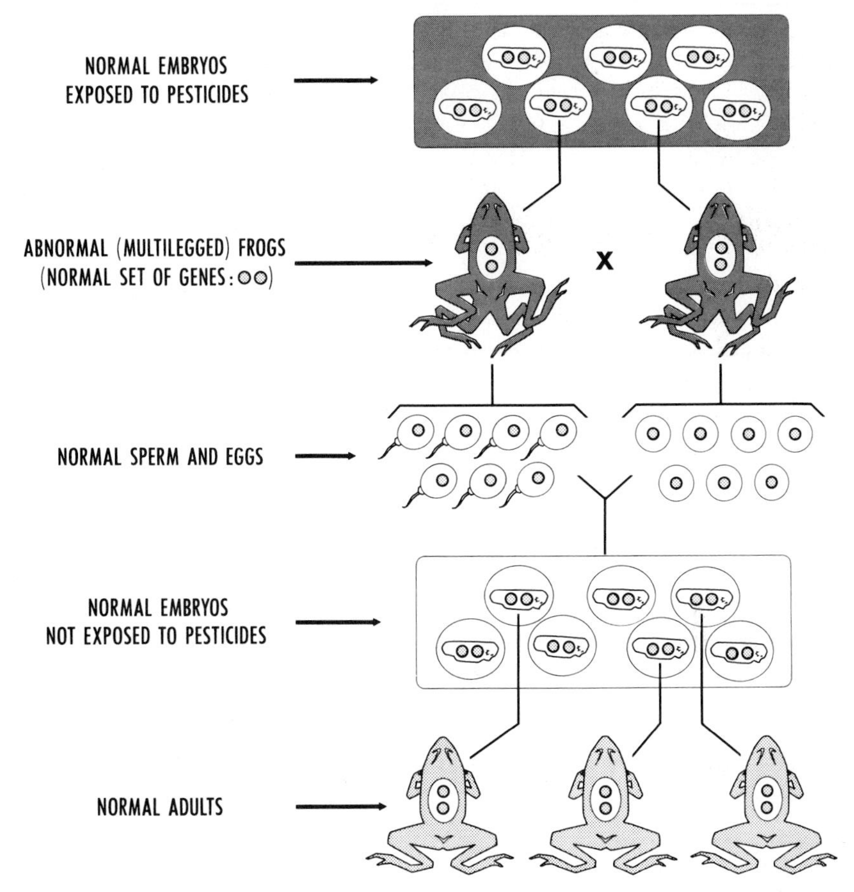

Figure 1.3 Test for the noninheritance of an environmentally induced trait. If the malformation of the hind limbs arose from without by the action of an adverse environmental agent (pesticide) and not from within by genetic change, then the abnormality would *not* be transmitted from parent to offspring. The multilegged parents, although visibly abnormal, are both genetically normal. Normal hind limbs are expected if, during development, the offspring are not exposed to the same injurious environmental agents as were the parents.

member of each pair coming from the male parent and the other from the female parent. Stated another way, the sperm and egg each has half the number of genes of the parents. Fertilization restores the original number and the paired condition (fig. 1.3).

The underlying assumption in the cross depicted in figure 1.3 is that the genes that influence the development of the hind limbs are normal ones. Of course, the multilegged parents themselves possess normal genes. It might seem strange that an abnormal character can result from a perfectly sound set of genes. However, normal genes cannot be expected to act normally under all environmental circumstances. A gene may be likened to a photographic negative. A perfect negative (normal gene) may produce an excellent or poor positive print (normal or abnormal trait) depending on such factors (environmental factors) as the quality or concentration of the chemical solutions used in preparing the print. The environment thus affects the *expression* of the negative (gene), but the negative (gene) itself remains unaffected throughout the making of the print (trait).

We shall learn more about the nature of genes and the mechanism of inheritance in the next chapter. For the moment, the important consideration is that *a given set of genes prescribes a potentiality for a trait and not the trait itself.* What is inherited is a potential capacity. The potential capacity will not become a developed capacity unless the appropriate environment is furnished. Genes always act within the conditioning framework of the environment.

Lamarckism

Few people would expect bodily deformities caused by harmful environmental factors to be inheritable. And yet, many persons believe that favorable or beneficial bodily changes acquired or developed during one's lifetime are transmitted to the offspring. As a familiar example, an athlete who exercised and developed large muscles would pass down his powerful muscular development to his children. This is the famous theory of the *inheritance of acquired characteristics,* or *Lamarckism* (after Jean Baptiste de Lamarck, a French naturalist of the late 1700s and early 1800s). The concept of Lamarckism has no foundation of factual evidence. We know, for example, that a woman who has altered her body by injections of silicone does not automatically pass the alterations on to her daughter. Circumcision is still necessary in the newborn male despite a rite that has been practiced for well over 4,000 years. It is sufficient to state that the results of countless laboratory experiments testing the possibility of the inheritance of acquired, or environmentally induced, bodily traits have been negative.

The notion of the inheritance of acquired characteristics became the cornerstone of Lamarck's comprehensive, although incorrect, explanation of evolution. Lamarck's theory is exemplified by a notable quotation from his book *Philosophie Zoologique* (1809):

> The giraffe lives in places where the ground is almost invariably parched and without grass. Obliged to browse upon trees it is continually forced to stretch upwards. This habit maintained over long periods of time by every individual of the race has resulted in the forelimbs becoming longer than the hind ones, and the neck so elongated that a giraffe can raise his head to a height of eighteen feet without taking his forelimbs off the ground.

Lamarck advocated that the organs of an animal became modified in appropriate fashion in direct response to a changing environment. The various organs became greatly improved through use or reduced to vestiges through disuse. Such bodily modifications, in some manner, could be transferred and impressed on the germ cells to affect future generations. Thus, the whale lost its hind limbs as the consequence of the inherited effects of disuse, and the giraffe developed its long neck through the inheritance of the effects of generations of continual stretching of the neck. Lamarck's views are unacceptable because we know of no mechanism that allows changes in the body to register themselves on the germ cells.

The foregoing considerations bear importantly on our bullfrog population. We have not been able to perform the critical breeding test to prove or disprove that the multilegged trait is a noninheritable modification. However, let us for the moment accept the thesis that the multilegged condition was environmentally induced and examine the implications of this view on the population as a whole. Outwardly, a striking change in the bullfrog population seems to have occurred. But in actuality the composition of the population has remained essentially unaltered, since the basic hereditary materials, the genes, were not affected. In other words, there was no change of evolutionary significance in the population. Evolution can occur *only* where there is inheritable variation.

Inheritable Variation

We shall now examine the possibility that the multilegged frogs were genetically abnormal. One or more of their genes may have been defective. Unlike teratogens, which damage an already conceived offspring and affect only a single generation, detrimental genes are in the germ plasm before conception and may lie dormant for several generations.

We shall postulate that the defective gene for the multilegged condition is *recessive* to the normal gene. That is to say, the expression of the defective gene is completely masked or suppressed by the normal (or *dominant*) gene when the two are present together in the same individual (fig. 1.4). Such an individual, normal in appearance but harboring the harmful recessive gene, is said to be a *carrier*. The recessive gene may be transmitted without any outward manifestation for several generations, continually being sheltered by its dominant partner. However, as seen in figure 1.4, the detrimental recessive gene ultimately becomes exposed when two carrier parents happen to mate. Those progeny that are endowed with two recessive genes, one from each parent, are malformed. On the average, one-fourth of the offspring will be multilegged.

Genetic defects transmitted by recessive genes are not at all unusual. Pertinent to the present discussion is an inherited syndrome of abnormalities in humans, known as the Ellis-van Creveld syndrome. Afflicted individuals are disproportionately dwarfed (short-limbed), have malformed hearts, and possess six fingers on each hand (fig. 1.5). The recessive gene that is responsible for this complex of defects is exceedingly rare. Yet, as shown by

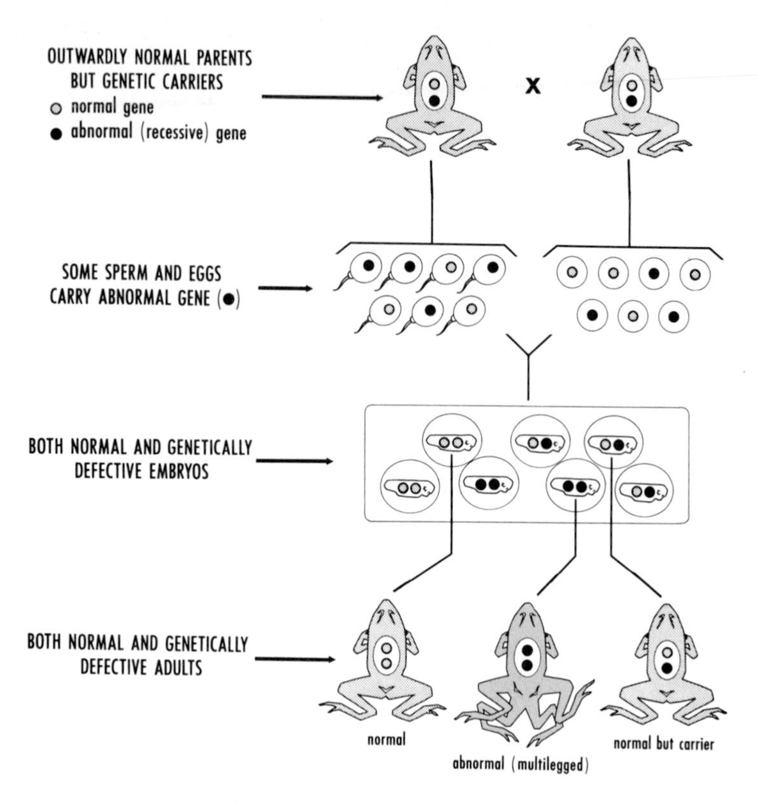

Figure 1.4 Mating of two "carrier" parents resulting in the emergence of multilegged offspring. The multilegged trait is assumed to be controlled by an abnormal recessive gene. Both parents are normal in appearance, but each carries the abnormal gene masked by the normal gene. The recessive gene manifests its detrimental effect when the offspring inherits one abnormal gene from each of its parents.

the geneticist Victor A. McKusick of Johns Hopkins University, the Ellis-van Creveld anomaly occurs with an exceptionally high incidence among the Amish people in Lancaster County, Pennsylvania. The defective recessive gene apparently was present in one member of the original Old Order Amish immigrants from Europe two centuries ago. For a few generations, the detrimental gene was passed down unobserved, masked by its normal partner gene. Since 1860, the Ellis-van Creveld deformity has appeared in at least 50 offspring. Ordinarily, it is uncommon for both members of a married couple to harbor the defective recessive gene. However, in the sober religious Amish community, marriages have been largely confined within members of the sect with a resulting high degree of consanguinity. Marriages of close relatives have tended to promote the meeting of two normal, but carrier, parents.

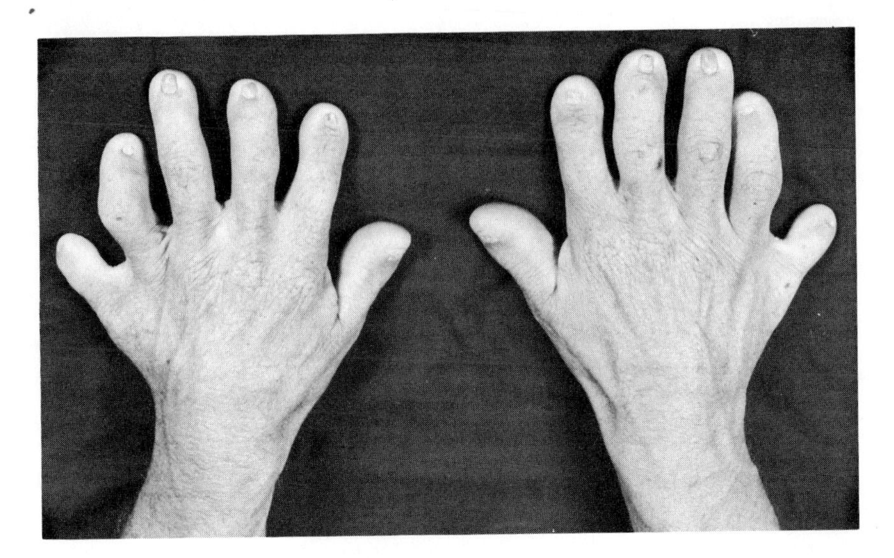

Figure 1.5 Six-digited hands, one of the manifestations of the Ellis-van Creveld syndrome in humans, a rare recessive deformity. Affected offspring generally come from two normal, but carrier, parents, each of which harbors the abnormal recessive gene. (Courtesy of Dr. Victor A. McKusick.)

We know very little about the past history or breeding structure of the particular bullfrog population in which the multilegged trait appeared with an exceptionally high frequency. We may suspect that all 350 or more multilegged frogs were derived from a single mating of two carrier parents. In contrast to humans, a single mated pair of frogs can produce well over 10,000 offspring. The similarity in age of the multilegged frogs found in nature adds weight to the supposition that these frogs are members of one generation, and probably of one mating.

Much of the preceding discussion on the bullfrog is admittedly speculative. However, one aspect is certain: *previously concealed harmful genes are brought to light through the mechanism of heredity.* A trait absent for many generations can suddenly appear without warning. Once a variant character expresses itself, its fate will be determined by the ability of the individual displaying the trait to survive and reproduce in its given environment.

It is difficult to imagine that the grotesquely shaped frogs could compete successfully with their normal kin. However, we shall never know whether or not the multilegged frogs were capable of contending with the severities of climatic or seasonal changes, or of successfully escaping their predators, or even of actively defending themselves. What we do know is that the multilegged frogs did not survive to reproductive age. Despite diligent searches by many interested investigators, no sexually mature abnormal frogs have been uncovered in the natural population. It seems that this unfavorable variant has been eliminated. Let us now explore this situation in light of Darwin's theory of natural selection.

Darwinian Evolution

Slightly more than a century ago, in 1859, Charles Robert Darwin (fig. 1.6) gave the biological world the master key that unlocked all previous perplexi-

Figure 1.6 Charles Darwin at the age of 31 (1840), four years after his famous voyage round the world as an unpaid naturalist aboard H. M. S. *Beagle*. (From a water color by George Richmond; courtesy of the American Museum of Natural History.)

ties about evolution. His revolutionary idea of natural selection can be compared only with Newton's law of gravitation and Einstein's theory of relativity. The concept of natural selection was set forth clearly and convincingly by Darwin in his monumental treatise *The Origin of Species.* This epoch-making book was the fruition of more than 20 years of meticulous accumulation and analysis of facts.

In 1831, Charles Darwin, then 22 years old and fresh from Cambridge University, accepted the unpaid post of naturalist aboard H.M.S. *Beagle,* a ship commissioned by the British Admiralty for a surveying voyage around the world. Although Darwin was an indifferent student at Cambridge, he did show an interest in the natural sciences. He was an earnest collector of beetles, enjoyed bird-watching and hunting, and was an amateur geologist.

It took the *Beagle* nearly five years—from 1831 to 1836—to circle the globe (fig. 1.7). When Darwin first embarked on the voyage, he did not dispute the dogma that every species of organism had come into being at the same time and had remained permanently unaltered. He shared the contemporary view that all organisms had been created about 4000 B.C.—more precisely, at 9:00 a.m. on Sunday, October 23, in 4004 B.C., according to the extraordinary pronouncement of Archbishop James Ussher in the 17th century. Darwin had in fact studied for the clergy at Cambridge University. But he was to make observations on the *Beagle's* voyage that he could not reconcile with accepted beliefs.

Darwin's quarters on the *Beagle* were cramped, and he took only a few books on board. One of them was the newly published first volume of Charles Lyell's *Principles of Geology,* a parting gift from his Cambridge mentor, John Henslow, professor of botany. Lyell rejected the prevailing belief that the earth's history had been characterized by successive episodes of creation and catastrophic destruction. He argued that the earth's mountains, valleys, rivers, and coastlines were shaped not by Noah's Flood but by the ordinary action of the rains, the winds, earthquakes, volcanoes, and other natural forces. Darwin was impressed by Lyell's emphasis on the great antiquity of the earth's rocks, and gradually came to perceive that the characteristics of organisms, as well as the face of the earth, could change over a vast span of time.

The living and extinct organisms that Darwin observed in the flat plains of the Argentine pampas and the Galápagos Islands sowed the seeds of Darwin's views on evolution. From old river beds in the Argentine pampas, he dug up bony remains of extinct mammals. One fossil finding was the massive *Toxodon,* whose appearance was likened to a hornless rhinoceros or a hippopotamus (fig. 1.8). Another fossil remain that attracted Darwin's attention was the skeleton of *Macrauchenia,* which he erroneously thought was clearly related to the camel because of the structure of the bones of its long neck. Other remarkable creatures were the huge *Pyrotherium,* resembling an elephant, and the light and graceful single-toed *Thoatherium,* rivaling the horse. The presence of *Thoatherium* testified that a horse had

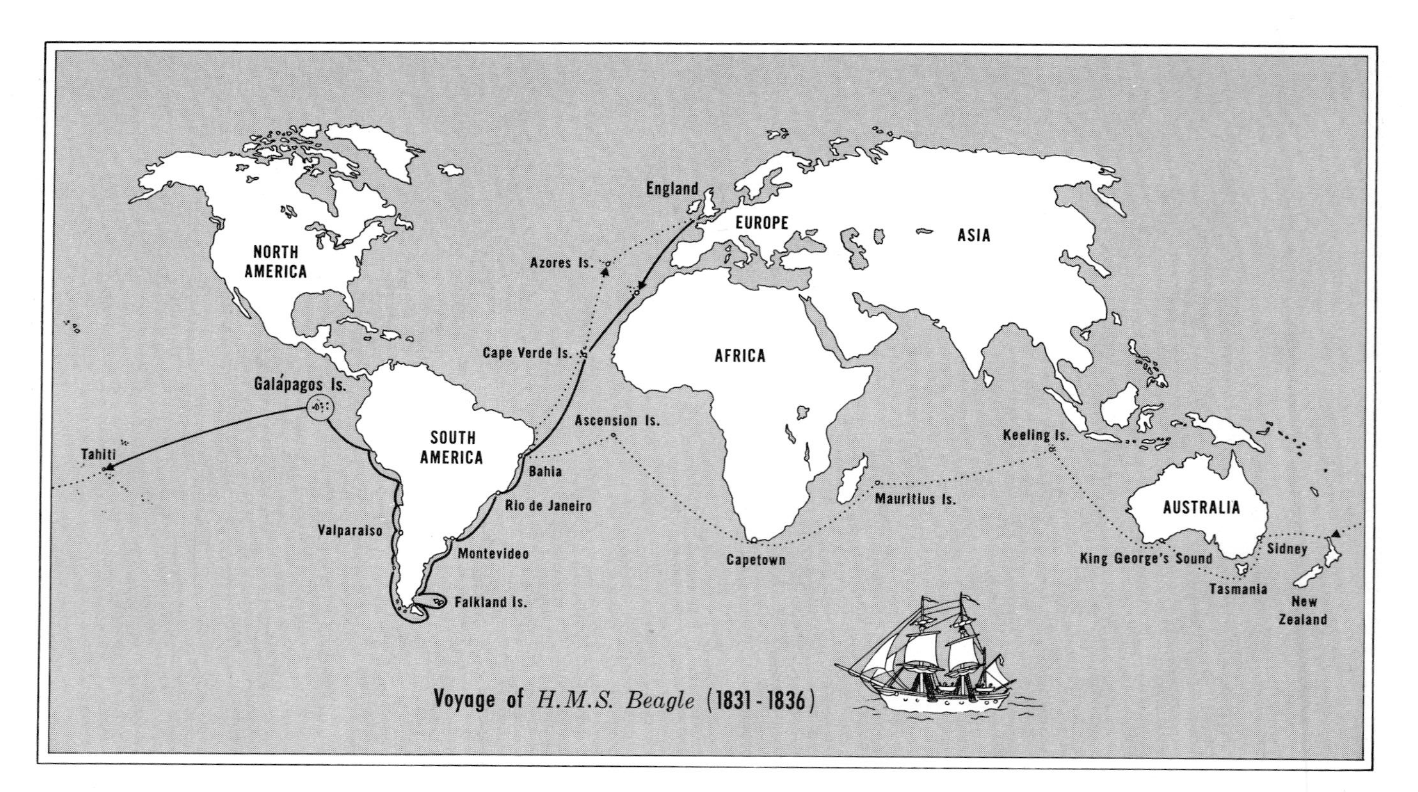

Figure 1.7 Five-year world voyage of H. M. S. *Beagle*. Darwin's observations on this voyage convinced him of the reality of evolution. Particularly impressive to him were the fossil skeletons of mammals unearthed in the Argentine pampas (see fig. 1.8) and the variety of tortoises and birds in the small group of volcanic islands, the Galápagos Islands (see fig. 12.3).

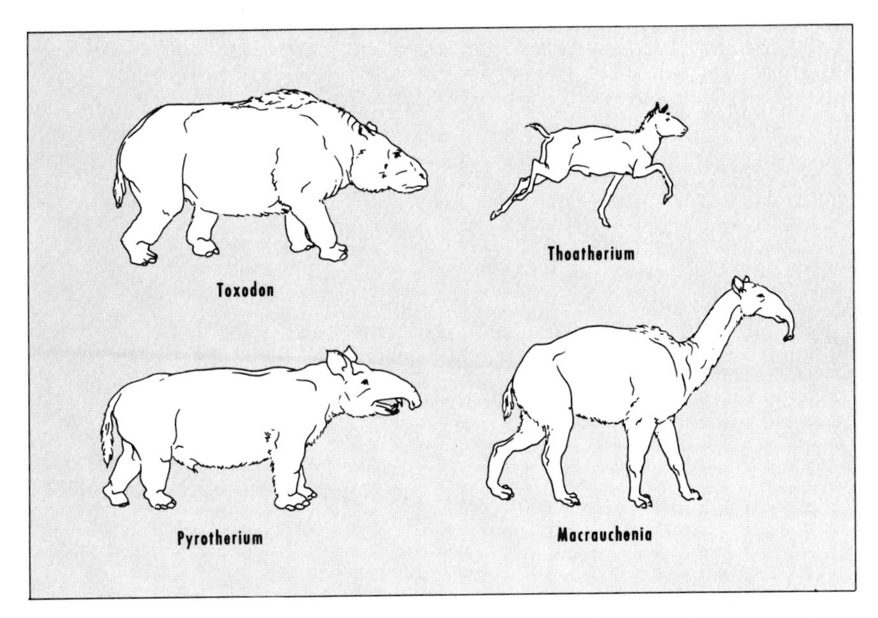

Figure 1.8 Curious hoofed mammals (ungulates) that flourished on the continent of South America some 60 to 70 million years ago, and have long since vanished from the scene. Bones of these great mammals were found by Darwin on the flat treeless plains of Argentina.

been among the ancient inhabitants of the continent. It was the Spanish settlers who reintroduced the modern horse, *Equus,* to the continent of South America in the 16th century. Darwin marveled that a native horse should have lived and disappeared in South America. This was one of the first indications that species gradually became modified with time, and that not all species survived through the ages.

When Darwin collected the remains of giant armadillolike and slothlike animals on an Argentine pampa, he pondered deeply on the fact that, although they clearly belonged to extinct forms, they were constructed on the same basic plan as the small living armadillos and sloths of the same region. This experience started him thinking of the fossil sequence of a given animal species through the ages and the causes of extinction. He wrote: "This wonderful relationship in the same continent between the dead and the living will, I do not doubt, hereafter throw more light on the appearance of organic beings on our earth, and their disappearance from it, than any other class of facts."

What Darwin had realized was that living species have ancestors. This fact is commonplace now, but it was a revelation then. On traveling from the north to the south of South America, Darwin observed that one species was replaced by similar, but slightly different, species. In the southern part of Argentina, Darwin caught a rare species of ostrich that was smaller and differently colored from the more northerly, common American ostrich, *Rhea americanus*. This rare species of bird was later named after him, *Rhea darwini*. It was scarcely imaginable to Darwin that several minor versions of a species would be created separately, one for each locality. It appeared to Darwin that species change not only in time but also with geographical distance. He later wrote: "It was evident that such facts could only be explained on the supposition that species gradually became modified; and the subject haunted me."

In the Galápagos Islands, Darwin's scientific curiosity was sharply prodded by the many distinctive forms of life. The Galápagos consist of an isolated cluster of islands of volcanic origin in the eastern Pacific, on the equator about 600 miles west of Ecuador. One of the most unusual animals is the giant land-dwelling tortoise, which may weigh as much as 500 pounds and attain an age of 200 to 250 years. The Spanish word for tortoise, *galápago*, gives the islands their name. Darwin noticed that the tortoises were clearly different from island to island, although the islands were only a few miles apart. In isolation, Darwin reasoned, each population had evolved its own distinctive features. Yet, all the island tortoises showed basic resemblances not only to each other but to relatively large tortoises on the adjacent mainland of South America. All this revealed to Darwin that the island tortoises shared a common ancestor with the mainland forms. The same was true of a group of small black birds, known today as *Darwin's finches*. Darwin observed that the finches were different on the various islands, yet they were obviously closely related to each other. Darwin reasoned that the finches were derived from an ancestral stock that originally populated the islands from the mainland. Apparently, a single ancestral group can give rise to several different varieties or species.

Mechanism of Evolution

When Darwin left England in 1831, he had accepted the established theory that different species had arisen all at once and remained unchanged. On his return, however, in October 1836, he was convinced of the truth of the idea of descent—that all organisms, including man, are modified descendants of previously existing forms of life. Darwin's observations on the nature and distribution of animals during his voyage assured him that evolution, or change, was a fact. We can see two stages in the development of Darwin's thoughts: the first was the realization of the fact of evolution; the second was an explanation of the process of evolution.

At home, Darwin began to assemble data relevant to the mechanism of evolution. He became intrigued by the extensive domestication of animals and plants brought about by man's conscious efforts. Throughout the ages, man has been a powerful agent in modifying wild species of animals and plants to suit his needs and whims. Through careful breeding programs, man determines which characteristics or qualities will be incorporated or discarded in his domesticated stocks. By conscious *selection,* man has perfected the toylike Shetland pony, the Great Dane dog, the sleek Arabian race horse, and vast numbers of cultivated crops and ornamental plants.

Many clearly different domestic varieties have evolved from a single species through man's efforts. As an example, we may consider the many varieties of domestic chickens (fig. 1.9), all of which are derived from a single wild species of red jungle fowl (*Gallus gallus*) present at one time in northern India. The red jungle fowl dates back to 2000 B.C. and has been so modified as to be nonexistent today. The variety of fowls perpetuated by fowl fanciers ranges from the flamboyant ceremonial cocks (the Japanese *Onaga-dori*) to the leghorns, bred especially to deposit spotless eggs.

In the late 1830s, Darwin attended the meetings of animal breeders and intently read their publications. Animal breeders were conversant with the variability in their pet animals, and dwelled on the technique of *artificial selection.* That is, the breeders selected and perpetuated those variant types

Figure 1.9 Evolution under domestication. A variety of domestic chickens have evolved from a single wild species (now extinct) as a result of man's continual practice of artificial selection.

that interested them or seemed useful to them. The breeders, however, had only vague notions as to the origin, or inheritance, of the variable traits.

Darwin acknowledged the unlimited variability in organisms, but was never able to explain satisfactorily how a variant trait was inherited. He and other naturalists were unaware of Gregor Mendel's contemporaneous discovery. Rather, Darwin followed Lamarck in assuming that biological variation is chiefly conditioned by direct influences of the environment on the organism. He believed that the changes induced by the environment became inherited; that bodily changes in the parents leave an impression or mark on their germ cells, or gametes. Darwin thought that all organs of the body sent contributions to the germ cells in the form of particles, or *gemmules*. These gemmules, or minute delegates of bodily parts, were supposedly discharged into the bloodstream and became ultimately concentrated in the gametes. In the next generation, each gemmule reproduced its particular bodily component. The theory was called *pangenesis,* or "origin from all," since all parental bodily cells were supposed to take part in the formation of the new individual.

Having explained the origin of variation (although incorrectly), Darwin wondered how artificial selection (a term familiar then only to animal breeders) could be carried on in nature. There was no breeder in nature to pick and choose. In 1837, Darwin wrote: "How selection could be applied to organisms living in a state of nature remained a mystery to me." Slightly more than a year later, Darwin found the solution. In his autobiography, Darwin explained: "In October 1838, that is fifteen months after I had begun my systematic enquiry, I happened to read for amusement *Malthus on Population* . . . at once it struck me that under these circumstances favourable variations would tend to be preserved and unfavourable ones destroyed. The result of this would be the formation of new species." The circumstances mentioned by Darwin were those associated with the assertions of an English clergyman, Thomas Robert Malthus, that population is necessarily limited by the means of subsistence.

Darwin's Natural Selection

Malthus' writings provided the germ for Darwin's thesis of *natural selection.* In his famous *An Essay on the Principle of Population,* Malthus expressed the view that the reproductive capacity of mankind far exceeds the food supply available to nourish an expanding human population. Men compete among themselves for the necessities of life. This unrelenting competition engenders vice, misery, war, and famine. It thus occurred to Darwin that competition exists among all living things. Darwin then envisioned that the "struggle for existence" might be the means by which the well-adapted individuals survive, and the ill-adjusted are eliminated. Darwin was the first to realize that perpetual selection existed in nature in the form of *natural selection.* In natural selection as contrasted to artificial selection, the animal breeder or horticulturist is replaced by the conditions of the environment that prevent the survival and reproduction of certain individuals. The process of natural selection occurs without a conscious plan or purpose. Natural selection was an entirely new concept, and Darwin was its proponent.

It was not until 1844 that Darwin developed his idea of natural selection in an essay, but not for publication. He showed the manuscript to the geologist Charles Lyell, who encouraged him to prepare a book. Darwin still took no steps toward publishing his views. It appears that Darwin might not have prepared his famous volume had not a fellow naturalist in the Dutch East Indies, Alfred Russel Wallace, independently conceived of the idea of natural selection. Wallace had spent many years exploring and collecting in South America and the East Indies. Wallace was also inspired by reading Malthus' essay, and the idea of natural selection came to him in a flash of insight during a sudden fit of malarial fever. In June of 1858, Wallace sent Darwin a brief essay on his views. The essay was entitled *On the Tendencies of Varieties to Depart Indefinitely from the Original Type.* With the receipt of this essay, Darwin was then induced to make a statement of his own with that of Wallace.

Wallace's essay and a portion of Darwin's manuscript, each containing remarkably similar views, were read simultaneously before the Linnaean Society in London on July 1, 1858. The joint reading of the papers stirred little interest. Darwin then labored for eight months to compress his voluminous notes into a single book, which he modestly called "only an Abstract." Wallace shares with Darwin the honor of establishing the mechanism by which evolution is brought about, but it was the monumental *The Origin of Species,* with its impressive weight of evidence and argument, that left its mark on mankind. The full title of Darwin's treatise was *On the Origin of Species by Means of Natural Selection, or the Preservation of Favoured Races in the Struggle for Life.* The first edition, some 1,500 copies, was sold out on the very day it appeared, November 24, 1859. The book was immediately both acidly attacked and effusively praised. Today, *The Origin of Species* remains the one book to be read by all serious students of nature.

The idea of evolution—that organisms change—did not originate with Darwin. There were many before him, notably Lamarck, Georges Buffon, and his grandfather, Erasmus Darwin, who recognized or intimated that animals and plants had not remained unchanged through time, but were continuously changing. Indeed, early Greek philosophers, who wondered about everything, speculated on the gradual progression of life from simple to complex. It was reserved for Darwin to remove the doctrine of evolution from the domain of speculation. Darwin's outstanding achievement was his discovery of the principle of natural selection. In showing *how* evolution occurs, Darwin convinced skeptics that evolution *does* occur.

As a whole, the principle of natural selection stems from three important observations and two deductions that logically follow from them. The first observation is that all living things tend to increase their numbers at a prolific rate. A single oyster may produce as many as 100 million eggs at one spawning; one tropical orchid may form well over 1 million seeds; and a single salmon can deposit 28 million eggs in one season. It is equally apparent (the second observation) that no one group of organisms swarms uncontrollably over the surface of the earth. In fact, the actual size of a given population of any particular organism remains relatively constant over long periods of time. If we accept these readily confirmable observations, the conclusion necessarily follows that not all individuals that are produced in any generation can survive. There is inescapably in nature an intense "struggle for existence."

Darwin's third observation was that individuals in a population are not alike but differ from one another in various features. That all living things vary is indisputable. Those individuals endowed with the most favorable variations, concluded Darwin, would have the best chance of surviving and passing their favorable characteristics on to their progeny. This differential survival, or "survival of the fittest," was termed *natural selection*. It was the British philosopher Herbert Spencer who proposed the expression "survival of the fittest," which Darwin accepted as equivalent to natural selection. In fact, Spencer suggested that Darwin had discovered not merely the laws of biological evolution but also those governing human societies.

Darwin presents the essence of his concept of natural selection in the introduction to the *Origin of Species,* as follows:

> As many more individuals of each species are born than can possibly survive; and as, consequently, there is a frequently recurring struggle for existence, it follows that any being, if it vary however slightly in any manner profitable to itself, under the complex and sometimes varying conditions of life, will have a better chance of surviving, and thus be *naturally selected.* From the strong principle of inheritance, any selected variety will tend to propagate its new and modified form.

Differential Reproduction

The survival of favorable variants is one facet of the Darwinian concept of natural selection. Equally important is the corollary that unfavorable variants do not survive and multiply. Consequently, natural selection necessarily embraces two aspects, as inseparable as the two faces of the same coin: the negative (elimination of the unfit) and the positive (perpetuation of the fit). In its negative role, natural selection serves as a conservative or stabilizing force, pruning out the aberrant forms from a population.

The superior, or fit, individuals are popularly extolled as those that emerge victoriously in brutal combat. Fitness has often been naively confused with physical, or even athletic, prowess. This glorification is traceable to such seductive catch phrases as the "struggle for existence" and the "survival of the fittest." But what does fitness actually signify?

The true gauge of fitness is not merely survival, but the organism's capacity to leave offspring. An individual must survive to reproduce, but not all individuals that survive do, or are able to, leave descendants. We have seen that the multilegged bullfrogs were not successful in propagating themselves. They failed to make a contribution to the next or succeeding generations. Therefore, they were unfit. Hence, an individual is biologically unfit if he leaves no progeny. He is also unfit if he does produce progeny, none of whom survives to maturity. The less spectacular normal-legged frogs did reproduce, and to the extent that they are represented by descendants in succeeding generations, they are the fittest. Fitness, therefore, is measured as reproductive effectiveness. Natural selection can thus be thought of as *differential reproduction,* rather than differential survival.

Evolution Defined

Any given generation is descended from only a small fraction of the previous generation. It should be evident that the genes transmitted by those individuals who most successfully reproduce will predominate in the next generation. Because of unequal reproductive capacities of individuals with different hereditary constitutions, the genetic characteristics of a population become altered each successive generation. This is a dynamic process that has occurred in the past, occurs today, and will continue to occur as long as inheritable variation and differing reproductive abilities exist. Under these circumstances, the composition of a population can never remain constant. This, then, is evolution—*changes in the genetic composition of a population with the passage of each generation.*

The outcome of the evolutionary process is adaptation of an organism to its environment. Many of the structural features of organisms are marvels of construction. It is, however, not at all remarkable that organisms possess particular characteristics that appear to be precisely and peculiarly suited to their needs. This is understandable because the individuals that leave the most descendants are most often those that are best equipped to cope with special environmental conditions. In other words, the more reproductively fit individuals tend to be those that are better adapted to the environment.

Throughout the ages, appropriate adaptive structures have arisen as the result of gradual changes in the hereditary endowment of a population. Admittedly, past events are not amenable to direct observation or experimental verification. There are no living eye-witnesses of very distant events. So, the process of evolution in the past must be inferred. Nevertheless, we may be confident that the same evolutionary forces we witness in operation today have guided evolution in the past. The basis for this confidence is shown in the chapters ahead.

Selected Readings

Darwin, C. 1967. *On the origin of species* (1859). New York: Atheneum Publishers.

Darwin, C. 1962. *The voyage of the Beagle* (1840). Garden City, N. Y.: Doubleday & Co.

Dobzhansky, T. 1956. *The biological basis of human freedom.* New York: Columbia University Press.

Eiseley, L. 1957. *The immense journey.* New York: Random House.

Eiseley, L. 1958. *Darwin's century.* Garden City, N. Y.: Doubleday & Co.

Gastonguay, P. R. 1974. *Evolution for everyone.* Indianapolis: Bobbs-Merrill Co.

Greene, J. C. 1961. *The death of Adam.* New York: The New American Library.

Hamilton, T. H. 1967. *Process and pattern in evolution.* New York: Macmillan Co.

Irvine, W. 1972. *Apes, angels, and victorians.* New York: McGraw-Hill Book Co.

Lerner, I. M. and Libby, W. J. 1976. *Heredity, evolution, and society.* San Francisco: W. H. Freeman and Co.

Merrell, D. J. 1962. *Evolution and genetics.* New York: Holt, Rinehart and Winston.

Moore, R. and editors of Time-Life Books. 1964. *Evolution.* New York: Time-Life Books.

Moorehead, A. 1969. *Darwin and the Beagle.* New York: Harper & Row.

Simpson, G. G. 1951. *The meaning of evolution.* New York: The New American Library.

Stebbins, G. L. 1971. *Processes of organic evolution.* Englewood Cliffs, N.J.: Prentice-Hall.

Wallace, B. and Srb, A. M. 1964. *Adaptation.* Englewood Cliffs, N.J.: Prentice-Hall.

2 Inheritable Variation

Darwin recognized that the process of evolution is inseparably linked to the mechanism of inheritance. But he could not explain satisfactorily how a given trait is transmitted from parent to offspring, nor could he account adequately for the sudden appearance of new traits. Unfortunately, Darwin was unaware of the great discovery in heredity made by a contemporary, the humble Austrian monk Gregor Johann Mendel.

Mendel's manuscript appeared in print in 1866. But his published paper was overlooked by most scientists of the day, including Darwin. The distinguished German botanist Karl von Nägeli corresponded with Mendel, but he surprisingly failed to grasp that Mendel had formulated the fundamental laws of inheritance. Mendel's valuable contribution lay ignored or unappreciated until 1900, sixteen years after his death. The rediscovery of Mendel's publication at the beginning of the 20th century ushered in the science of genetics and ultimately led to our current understanding of inheritable variation.

Mendel's Law of Segregation

Mendel's profound inference was that traits are passed on from parent to offspring through the gametes, in specific discrete factors, or units. The individual units do not blend and do not contaminate one another. The hereditary units of the parents can therefore be reassorted in varied combinations in different individuals at each generation. Today, Mendel's hereditary units are called *genes*. The almost infinite diversity that exists among individuals is attributable largely to the shuffling of tens of thousands of discrete genes that occurs in sexual reproduction.

The units of heredity, or genes, typically occur in pairs in the fertilized egg. The offspring inherits one gene of each pair from the female parent and one of each pair from the male parent. In the most elementary case of inheritance, the trait is governed by a single pair of genes. Albinism in man is an example of a trait under the simple control of one pair of genes.

The word "albino" is derived from the Latin *albus,* meaning white, and refers to the paucity or absence of pigment (melanin) in the skin, hair, and eyes. The skin is often very light ("milk-white"), and the hair whitish-yellow. The eyes appear pinkish because the red blood vessels give the otherwise colorless iris a rosy cast. Albinos have poor vision and are acutely sensitive to sunlight. They are especially prone to skin cancer. In some cultures, as among the natives of the Archipelago Coral Islands, albinos are believed to have a mesmeric aura (fig. 2.1).

Albinism results from a defective gene. Specifically, the condition arises only when the individual is endowed with two defective genes, one from each parent. Remember that the affected person must have received the defective gene for albinism from *both* parents.

Figure 2.1 Albino person showing the complete absence of pigment in the eyes, hair, and skin. Natives of the Archipelago Coral Islands (100 miles east of the New Guinea mainland) are said to take care of their "pure white" relatives who can hardly see in the glare of the tropical sun. (Courtesy of Wide World Photos.)

As seen in figure 2.2, there are nine possible marriages with respect to the pair of genes conditioning the presence or absence of melanin. If both parents are perfectly normal—that is, each possesses a pair of normal genes—then all offspring will be normally pigmented (cross 1). When both parents are albino and accordingly each has a pair of defective genes, all offspring will be afflicted with albinism (cross 2). The persons involved in either of these two crosses are said to be *homozygous,* since they each possess a pair of similar, or like, genes. The normal individuals in cross 1 are homozygous for the normal gene, and the albino individuals in cross 2 are homozygous for the defective gene.

Looking at cross 3, we notice that one of the parents (the mother) carries an unlike pair of genes: one member of the pair is normal; its partner gene is defective. This parent, possessing unlike genes, is *heterozygous.* This parent has normal coloration, which leads us to deduce that the expression of the defective gene is completely masked or suppressed by the normal gene. In genetical parlance, we say that the normal gene is *dominant* over its partner, the *recessive* gene. The dominant gene is customarily symbolized by a capital letter (in our case, *A*); the recessive gene, by a corresponding small letter(*a*). The heterozygous parent is depicted as *Aa,* and is referred to as *heterozygote* or a *carrier.*

When the heterozygote (*Aa*) marries a dominant homozygote (*AA*), all the offspring will be outwardly normal, but half the offspring will be carriers (*Aa*) like one of the parents (cross 3). The ratio of *AA* to *Aa* offspring may be expressed as 1/2 *AA* : 1/2 *Aa* or simply 1 *AA* : 1*Aa.* Cross 4 differs from cross 3 only in that the male parent is the carrier (*Aa*) and the female parent is the dominant homozygote (*AA*). The outcome is the same: all children are normal in appearance; but half of them are carriers. Thus, with respect to albinism, the males and females are equally likely to have or to transmit the trait. (There are traits, such as hemophilia and red-green color-blindness, in which transmission is influenced by the sex of the parents.)

All offspring will be carriers from the marriage (crosses 5 and 6) of a dominant homozygote (*AA*) and an albino parent (*aa*). Examining crosses 5 and 6 (fig. 2.2), we again notice that males and females are equally affected. The marriage of a heterozygous person (*Aa*) and an albino (*aa*) gives rise to two types of progeny, in equal numbers (crosses 7 and 8). Half the offspring are normal but carriers (*Aa*); the remaining half are albino (*aa*).

A cardinal fact of inheritance is that each gamete (egg or sperm) contains *one, and only one,* member of a pair of genes. Thus, when the egg and sperm unite, the two genes for each character are brought together in the new individual. During the production of the gametes, the members of a pair of genes separate, or *segregate,* from each other. A given gamete can carry *A* or *a,* but not both. This fundamental concept that only one member of any pair of genes in a parent is transmitted to each offspring is Mendel's first law—the *law of segregation.*

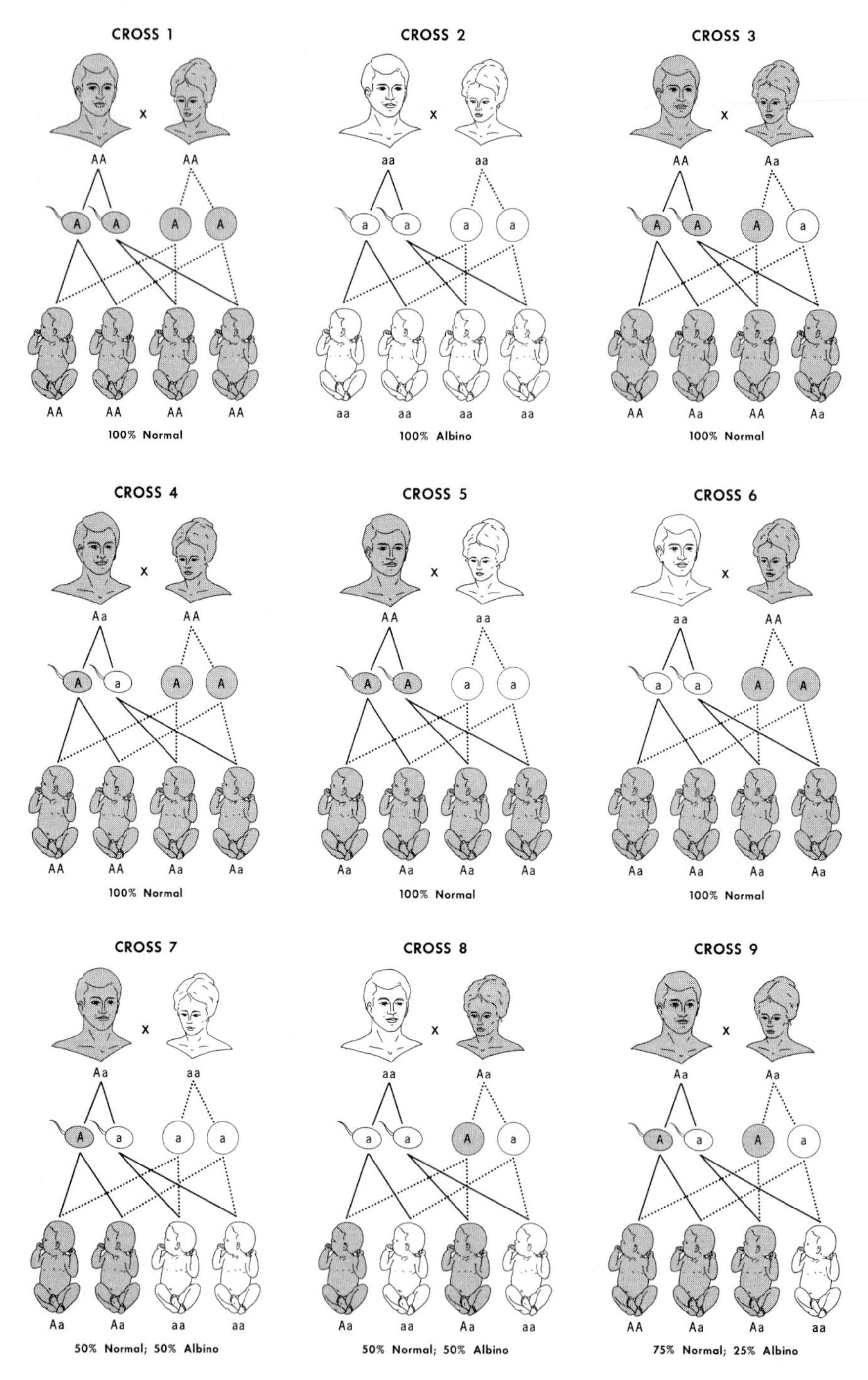

Figure 2.2 The segregation of a single pair of genes. The nine possible marriages with respect to the recessive trait for albinism show the operation of Mendel's law of segregation.

The outcome of the last of the crosses (no. 9) in figure 2.2 is comprehensible in terms of Mendel's law of segregation. Both parents are normal in appearance, but each carries the defective gene (*a*) masked by the normal gene (*A*). The male parent (*Aa*) produces two kinds of sperm cells: half the sperm carry the *A* gene; half carry the *a* allele. The same two kinds occur in equal proportions among the egg cells. Each kind of sperm has an equal chance of meeting each kind of egg. The random meeting of gametes leads to both normal and albino offspring. One-fourth of the progeny are normal and completely devoid of the recessive gene (*AA*); one-half are normal but carriers like the parents (*Aa*); and the remaining one-fourth exhibit the albino anomaly (*aa*). The probabilities of occurrence are the same as those illustrated by the tossing of a coin. If two coins are repeatedly tossed at the same time, the result to be expected on the basis of pure chance is that the combination "head and head" will fall in one-fourth of the cases, the combination "head and tail" in one-half of them, and the combination "tail and tail" in the remaining one-fourth. The ratio of genetic constitutions among the offspring may be expressed as 1/4 *AA* : 1/2 *Aa* : 1/4 *aa,* or 1 *AA* : 2 *Aa* : 1 *aa.*

Geneticists constantly have to make a distinction between the external, or observable, appearance of the organism and its internal hereditary constitution. They call the former the *phenotype* (visible type) and the latter the *genotype* (hereditary type). In our last cross (no. 9), there are three different genotypes: *AA, Aa,* and *aa.* The dominant homozygote (*AA*) cannot be distinguished by inspection from the heterozygote (*Aa*). Thus, *AA* and *Aa* genotypes have the same phenotype. Accordingly, on the basis of phenotype alone, the progeny ratio is 3/4 normal : 1/4 albino, or 3 normal : 1 albino.

If we disregard the sexes of the parents, the nine crosses in figure 2.2 become reducible to six possible types of matings. The six kinds of matings involving a single pair of genes are set forth in table 2.1.

Application of Mendelian Principles

Our interpretation of the mode of inheritance of the multilegged trait in the bullfrog was based on Mendel's principles. In the preceding chapter, we postulated a cross between two normal, but heterozygous, frogs. The expectation for multilegged offspring from two heterozygous parents is 25 percent, in keeping with Mendel's 3:1 ratio, This cross, using appropriate genetic symbols, is shown graphically by "checkerboard" analysis in figure 2.3. The convenient checkerboard method of analysis was devised by the British geneticist R. C. Punnett. The eggs and sperm are listed separately on two different sides of the checkerboard, and each square represents an offspring that arises from the union of a given egg cell and a given sperm cell. The "reading" of the squares discloses the classical Mendelian phenotypic ratio of 3:1.

Table 2.1
Simple Recessive Mendelian Inheritance, Involving a Single Pair of Genes (Normal Pigmentation vs. Albinism)

Mating Types		Gametes				Offspring	
Genotypes	Phenotypes	First Parent 50%	First Parent 50%	Second Parent 50%	Second Parent 50%	Genotypes	Phenotypes
AA x *AA*	Normal x Normal	*A*	*A*	*A*	*A*	100% *AA*	100% Normal
AA x *Aa*	Normal x Normal	*A*	*A*	*A*	*a*	50% *AA* 50% *Aa*	100% Normal
Aa x *Aa*	Normal x Normal	*A*	*a*	*A*	*a*	25% *AA* 50% *Aa* 25% *aa*	75% Normal 25% Albino
AA x *aa*	Normal x Albino	*A*	*A*	*a*	*a*	100% *Aa*	100% Normal
Aa x *aa*	Normal x Albino	*A*	*a*	*a*	*a*	50% *Aa* 50% *aa*	50% Normal 50% Albino
aa x *aa*	Albino x Albino	*a*	*a*	*a*	*a*	100% *aa*	100% Albino

It should be understood that the 3:1 phenotypic ratio resulting from the mating of two heterozygous persons, or the 1:1 ratio from the mating of a heterozygote and a recessive, are expectations based on probability, and not invariable outcomes. The production of large numbers of offspring increases the probability of obtaining, for example, the 1:1 progeny ratio, just as many tosses of a coin improve the chances of approximating the 1 head : 1 tail ratio. If a coin is tossed only two times, a head on the first toss is not invariably followed by a tail on the second toss. In like manner, if only two offspring are produced from a marriage of a heterozygote and a recessive parent, it should not be thought that one normal offspring is always accompanied by one recessive offspring. With small numbers of progeny, as is characteristic of man, any ratio might arise in a given family.

Stated another way, the 3:1 and 1:1 ratios reveal the risk or odds of a given child having the particular trait involved. For example, if the first child of two heterozygous parents is an albino, the odds that the second child will be an albino remain 1 out of 4. These odds hold for each subsequent child, irrespective of the number of previously affected children. Each conception is an entirely independent event. If this still appears puzzling, consider once again the tossing of a coin. The first time a coin is tossed, the chance of obtaining either a head or a tail is 1 in 2. Whether the toss is repeated immediately or nine months later, the chance of the coin falling head or tail is still 1 in 2, or 50 percent.

NORMAL (Heterozygous) PARENTS

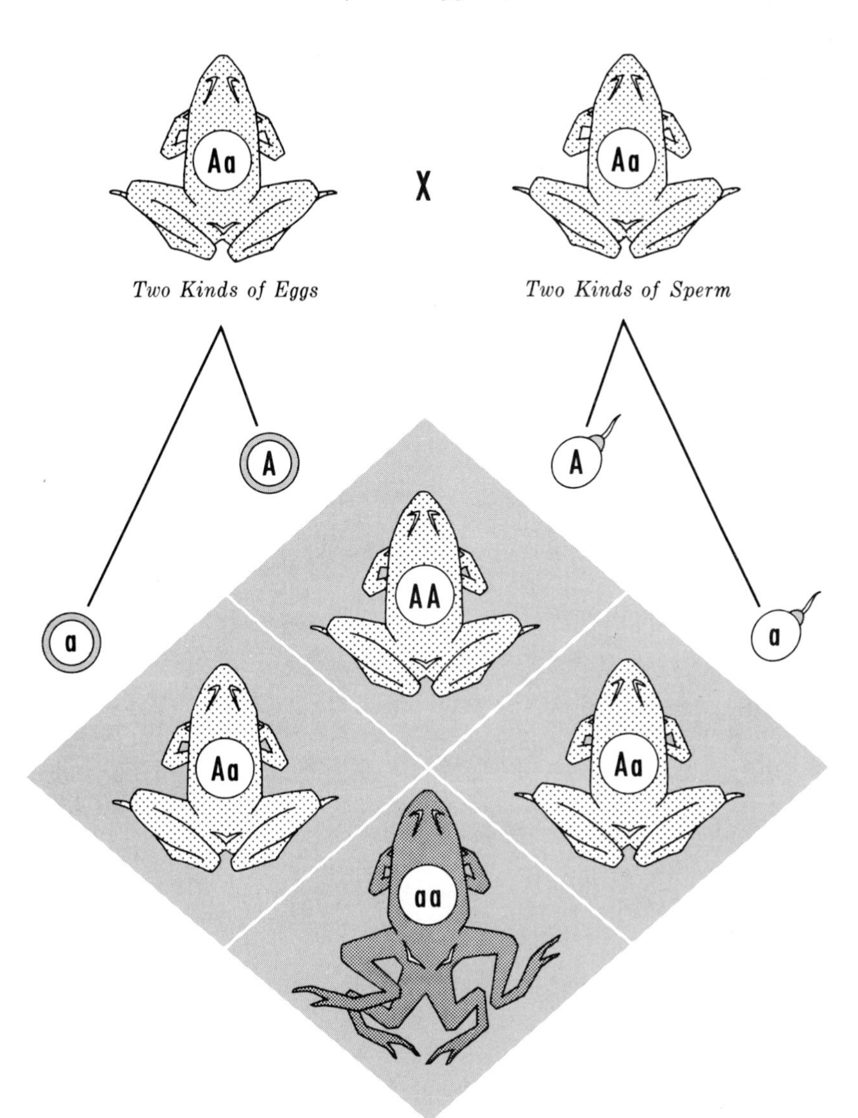

3/4 NORMAL : 1/4 MULTILEGGED

Figure 2.3 Cross of two heterozygous carrier frogs. One-fourth of the offspring are normal and completely devoid of the recessive gene (*AA*); one-half are normal but carriers, like the parents (*Aa*); and the remaining one-fourth exhibit the multilegged anomaly (*aa*).

Chromosomal Basis of Heredity

Each organism has a definite number of chromosomes, ranging in various species from 2 to more than 200. Man, for example, has 46 microscopically visible, threadlike chromosomes in the nucleus of every body cell (fig. 2.4).

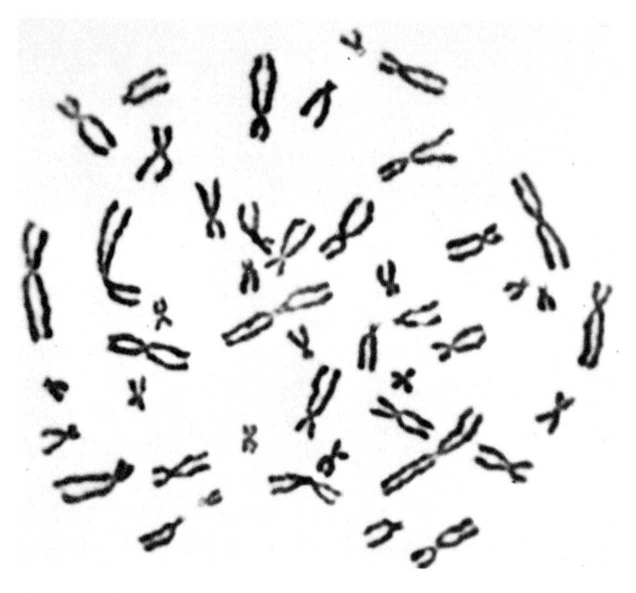

Figure 2.4 Normal human complement of 46 chromosomes, prepared from a culture of blood cells of a male. Each chromosome consists of two strands. The chromosomes occur in pairs; one member of each chromosome pair is derived from the mother and the other is derived from the father. Thus, each chromosome has a homologous ("like") mate in the complement.

Not only are the numbers of chromosomes constant for a given species, but the shapes and sizes of the chromosomes in a given species differ enough for us to recognize particular chromosomes and to distinguish one kind from the other. An important consideration is that each kind of chromosome in a zygote is present in duplicate. Stated another way, the chromosomes occur in pairs; man has 23 pairs. The significance of the double, or paired, set of chromosomes, cannot be overstated: *each parent contributes one member of each pair of chromosomes to the offspring.* The behavior of chromosomes parallels the behavior of genes. Mendel's findings revealed that the genes occur in pairs, and that the members of each pair of genes separate from each other in the production of the gametes. Evidently then, we may deduce that the genes are located in the chromosomes. Extensive studies have provided strong confirmation that the chromosomes do indeed carry the genes.

Two chromosomes that form a pair are *homologous.* A special kind of division guarantees that a gamete contains one member of homologous pairs of chromosomes. This special process is called *meiosis*. Meiosis is a lessening or reducing process, referring specifically to a reduction in the number of chromosomes. In man, for example, during formation of the gametes, the number of chromosomes is reduced from 46 to 23.

Figure 2.5 shows the separation of the pair of homologous chromosomes during meiosis of the heterozygote carrying the *A* and *a* genes. The gene *A* occupies a particular site, or *locus,* in one of the chromosomes. The alternative gene, *a,* occurs at the identical locus in the other homologous chromosome. The alternative forms of a given gene, occupying a given locus, are termed *alleles*. Thus, *A* is an allele of *a* (or, *A* is allelic to *a*). Now, we had earlier seen that the heterozygote (*Aa*) produces two types of gametes, *A* and *a,* in equal numbers. It should be evident from figure 2.5 that the gametic ratio of 1 *A* : 1 *a* is the consequence of the separation of the two homologous chromosomes during meiosis, one chromosome containing the *A* gene and its homologue carrying the *a* allele. The behavior of chromosomes also explains another feature of Mendel's results—the independent assortment of traits.

Mendel's Law of Independent Assortment

The fruit fly (*Drosophila melanogaster*), approximately one-fourth the size of a house fly, has long been a favorite subject for genetical research. Numerous variations have been found, affecting all parts of the body. For example, the wings may be reduced in length to small vestiges, the result of action of a recessive, *vestigial* gene. Two heterozygous normal-winged flies, each harboring the recessive gene for vestigial wings, can produce two types of offspring: normal-winged and vestigial-winged, in a ratio of three-to-one. Now suppose that these same parents have a second concealed recessive gene, namely, *ebony*. This recessive gene, when homozygous, modifies the normal gray color of the body to black, or ebony.

Two normal-winged, gray-bodied parents, each heterozygous (*NnGg*), can give rise to four types of offspring phenotypically, as seen in figure 2.6. Two of the phenotypes are like the original parents, and two are new combinations (normal-winged, ebony and vestigial-winged, gray). The four phenotypes appear in a ratio of 9:3:3:1. For such a ratio to be obtained, each parent (*NnGg*) must have produced four kinds of gametes in equal proportions: *NG, Ng, nG,* and *ng*. Stated another way, the segregation of the members of one pair of genes must have occurred independently of the segregation of the members of the other gene pair during gamete formation. Thus, 50 percent of the gametes received *N,* and of these same gametes, half obtained *G* and the other half *g*. Accordingly, 25 percent of the gametes were *NG* and 25 percent were *Ng*. Likewise, 50 percent of the gametes carried *n,* of which half contained as well *G* and half *g* (or 25 percent *nG*

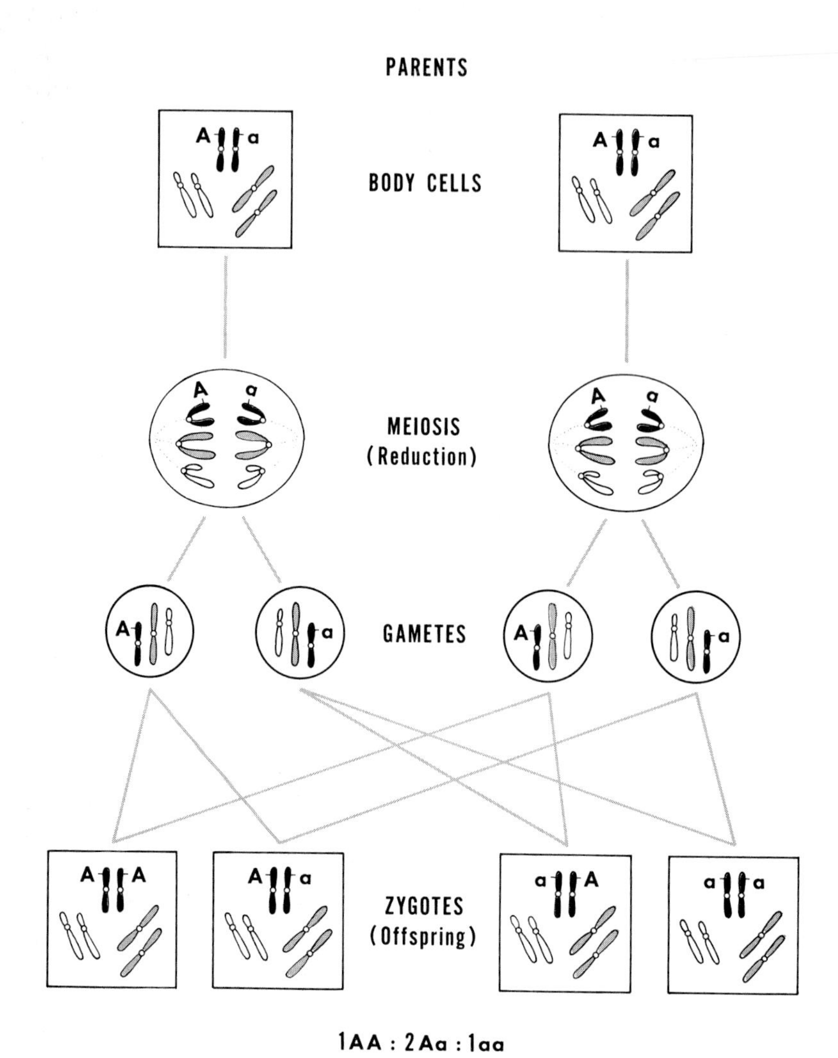

Figure 2.5 Chromosomal basis of inheritance. When one chromosome of a pair carries a given gene (*A*, in this case), and its homologue carries the alternative form of the gene (*a*), then meiosis results in the production of two distinct kinds of gametes in equal proportions, *A* and *a*.

and 25 percent *ng*). The four kinds of egg cells and the four kinds of sperm cells can unite in 16 possible ways, as shown graphically in the "checkerboard" in figure 2.6.

The above cross illustrates Mendel's *law of independent assortment,* which states that one trait (one gene pair) segregates independently of an-

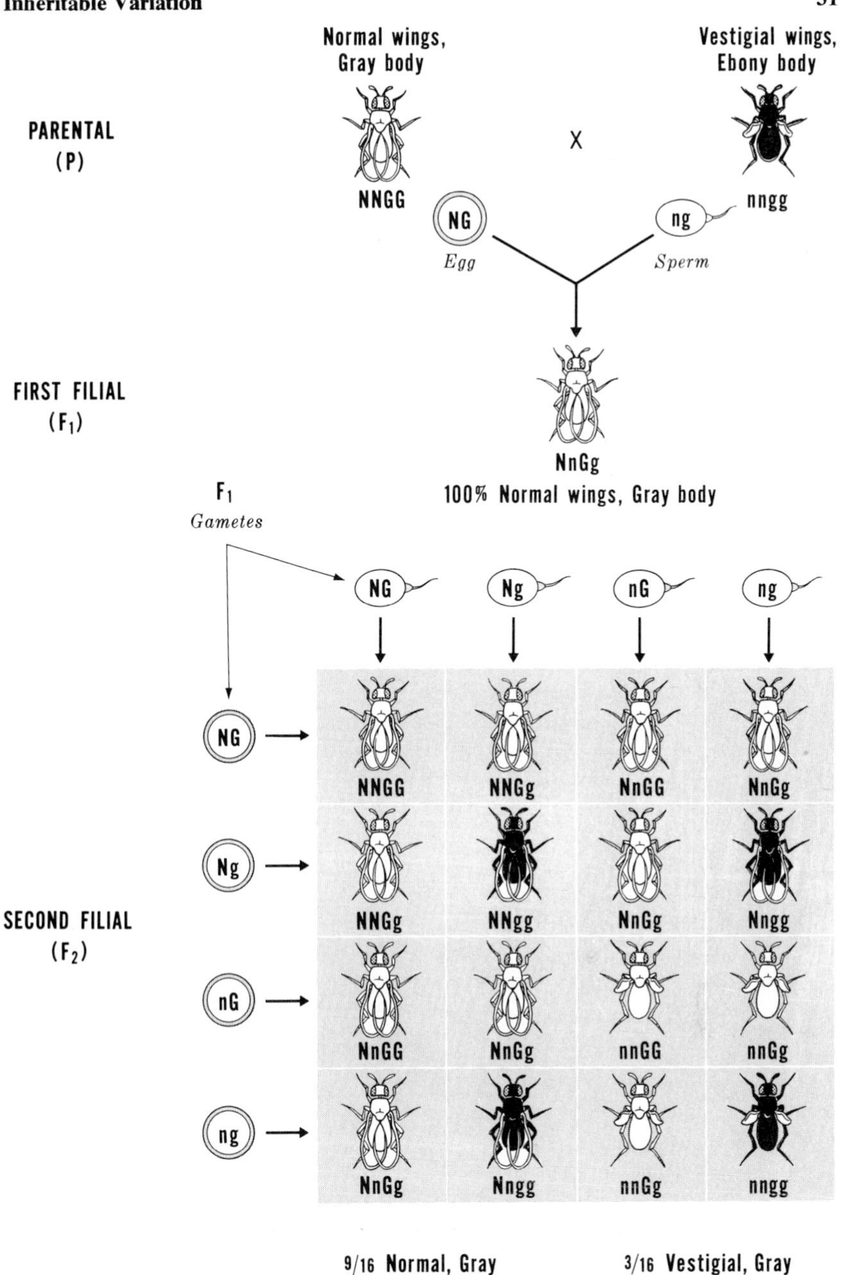

Figure 2.6 Mendel's law of independent assortment, as revealed in the inheritance of two pairs of characters in the fruit fly. A pure normal-winged fly with gray body mated with a vestigial-winged, ebony-bodied fly produces all normal-winged, gray-bodied flies in the first generation (F_1). When these F_1 flies are inbred, a second generation (F_2) is produced that displays a phenotypic ratio of 9:3:3:1.

other trait (another gene pair). The separation of the *Nn* pair of genes and and the simultaneous independent separation of the *Gg* pair of genes occurs because the two pairs of genes are located in two different pairs of chromosomes. Once again, the parallelism in the behavior of genes and chromosomes is striking.

Three or More Gene Pairs

When individuals differing in three characteristics are crossed, the situation becomes more complex, but the principle of independent assortment still holds. An individual heterozygous for three independently assorting pairs of alleles (for example, *Aa, Bb, and Cc*) produces eight different types of gametes in equal numbers, as illustrated in figure 2.7. Random union among

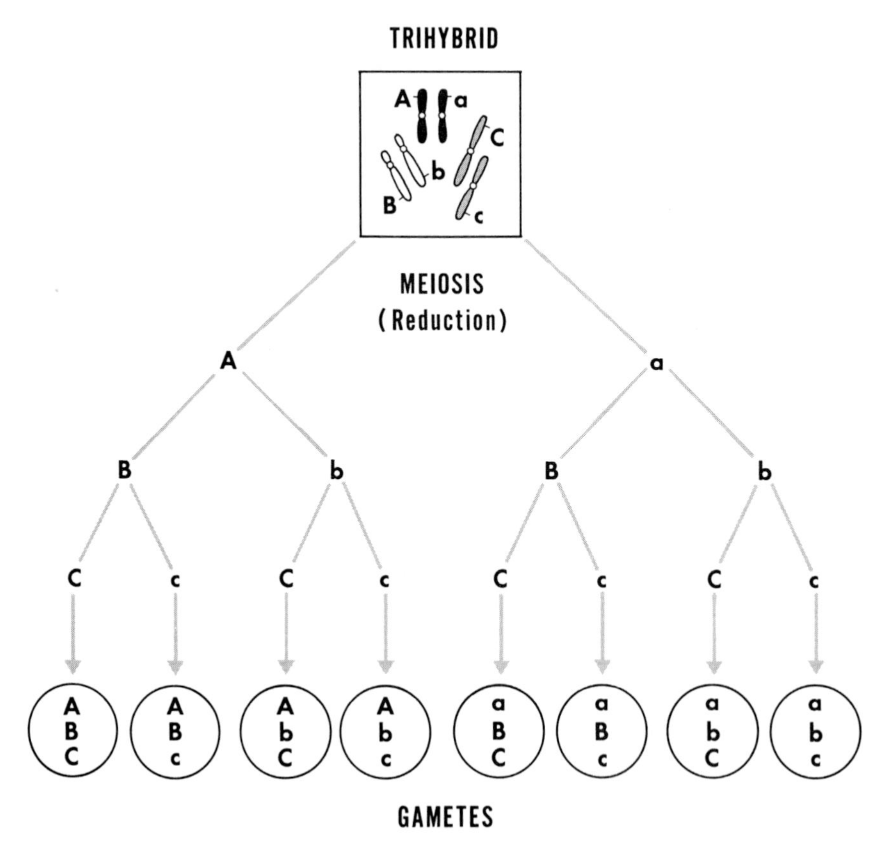

Figure 2.7 Gamete formation in an individual heterozygous for three pairs of genes ("trihybrid"). The members of the three different pairs of genes assort independently of each other when gametes are formed. The separation of the *Aa* pair, *Bb* pair, and *Cc* pair, respectively, can be treated as independent events. The outcome is eight genetically different gametes.

these eight kinds of gametes yields 64 equally possible combinations. Stated another way, Punnett's checkerboard method of analysis would have 64 (8 x 8) squares. An analysis of the offspring reveals that, among the 64 possible combinations, there are only eight visibly different phenotypes. You may wish to verify that the phenotypic ratio is 27:9:9:9:3:3:3:1.

The use of the Punnett square for three or more independently assorting genes is cumbersome. Here we will introduce one of the cardinal rules of probability: *The chance that two or more independent events will occur together is the product of their chances of occurring separately.* As an example, what proportion of the offspring of the cross *AaBbCc* x *AaBbCc* would be expected to have the genotype *AaBBcc?* The individual computations are as follows:

1. The chance that an individual will be *Aa* = 1/2.
2. The chance that an individual will be *BB* = 1/4.
3. The chance that an individual will be *cc* = 1/4.

Assuming independence between these three pairs of alternatives, the chance of the three combining together (*AaBBcc*) is the product of their separate probabilities, as follows:

$$1/2 \times 1/4 \times 1/4 = 1/32.$$

Significance of Independent Assortment

The Mendelian principles permit a genuine appreciation of the source of inheritable variation in natural populations of organisms. An impressively large array of different kinds of individuals can arise from the segregation and reassortment of independent pairs of genes. Figure 2.8 considers a few of the independently assorting traits in the fruit fly. It is seen that the number of visibly different classes (phenotypes) of offspring is doubled by each additional heterozygous pair of independently assorting genes. Each additional heterozygous gene pair multiplies the number of visibly different comminations by a factor of two. Thus, if the parents are each heterozygous for ten pairs of genes (when dominance is complete), the number of different phenotypes among the progeny becomes 2^{10}, or 1,024.

Other mathematical regularities are evident as we increase the number of heterozygous pairs (table 2.2). The number of different genotypes among the progeny is a multiple of the base 3, that is, 3^n. Moreover, as the number of heterozygous traits increases, the chance of recovering one of the homozygous classes becomes progressively less. Thus, when a single heterozygous pair (*Aa*) is involved, one in four will be either *AA* or *aa*. When four heterozygous traits are involved, only one in 256 will be either homozygous dominant or recessive for all traits. Evidently, no single genetic constitution is ever likely to be exactly duplicated in a person (save in identical twins). Each individual is genetically unique.

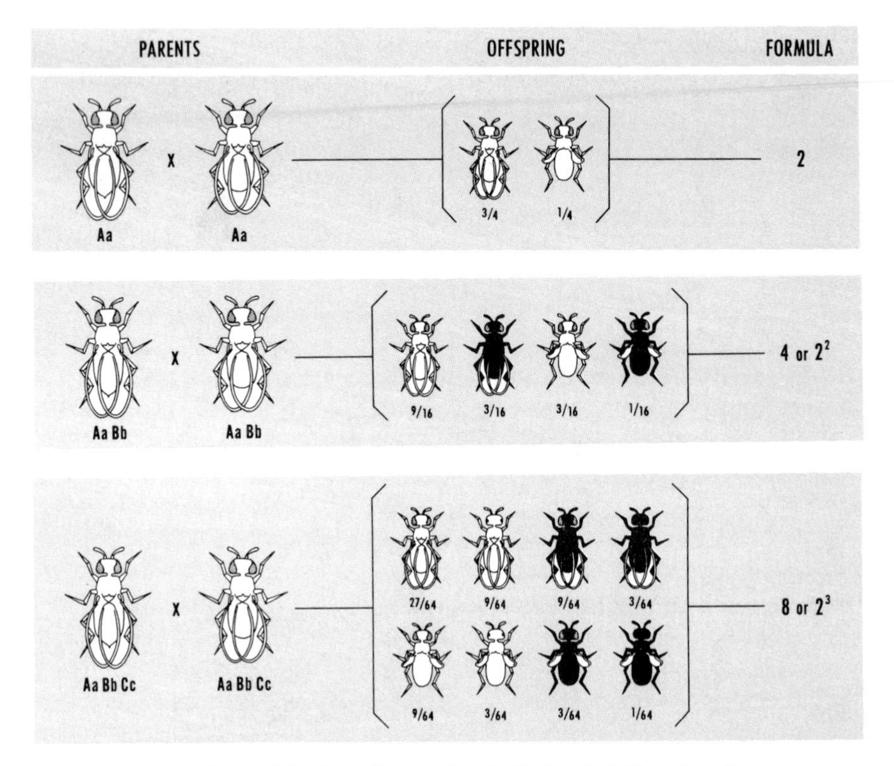

Figure 2.8 Inheritance of traits in the fruit fly, when the parents are each heterozygous for one (*Aa*), two (*AaBb*), and three (*AaBbCc*) independently assorting traits. The gene for long wings (*A*) is dominant over the gene for vestigial wings (*a*); gray body (*B*) is dominant over ebony (*b*); and normal eyes (*C*) is dominant over the eyeless condition (*c*). The number of visibly different classes (phenotypes) of offspring increases progressively with each additional pair of traits. The generalized formula for determining the number of different phenotypes among the offspring (when dominance is complete) is 2^n, where n stands for the number of different heterozygous pairs of genes.

The process of independent combination of genes may be compared, on a very modest scale, to the shuffling and dealing of playing cards. One pack of 52 cards can yield a large variety of hands. And, just as in poker a full house is far superior to three of a kind, so in organisms certain combinations of genes confer a greater reproductive advantage to its bearer than other combinations. Natural selection favors the more reproductive genotypes.

Table 2.2
Characteristics of Crosses Involving Various Pairs of Independently Assorting Genes

Number of heterozygous pairs involved in the cross	Number of different types of gametes produced by the heterozygote	Number of visibly different progeny (that is, phenotypes) *	Number of genotypically different combinations (that is, genotypes)	Chance of recovering an individual that is either homozygous dominant or recessive for all traits
1	2	2	3	1 in 4
2	4	4	9	1 in 16
3	8	8	27	1 in 64
4	16	16	81	1 in 256
10	1,024	1,024	59,049	1 in 1,048,576
n	2^n	2^n	3^n	1 in 4^n

* Assumes complete dominance in each pair.

Selected Readings

Beadle, G. W. and Beadle, M. 1968. *The language of life.* Garden City, N.Y.: Doubleday & Co.

Gardner, E. J. 1968. *Principles of genetics.* New York: John C. Wiley & Sons.

Iltis, H. 1932. *Life of Mendel.* New York: W. W. Norton & Co.

McKusick, V. A. 1969. *Human genetics.* Englewood Cliffs, N.J.: Prentice-Hall.

Mendel, G. 1948. *Experiments in plant hybridization.* Cambridge, Mass.: Harvard University Press. (Translation of Gregor Mendel's original paper first published in 1866.)

Moody, A. 1967. *Genetics of man.* New York: W. W. Norton & Co.

Moore, J. A. 1972. *Heredity and development.* New York: Oxford University Press.

Peters, J. A. 1959. *Classic papers in genetics.* Englewood Cliffs, N.J.: Prentice-Hall.

Nagle, J. J. 1974. *Heredity and human affairs.* Saint Louis: C. V. Mosby Co.

Scheinfeld, A. 1965. *Your heredity and environment.* Philadelphia: J. B. Lippincott Co.

Srb, A. M., Owen, R. D., and Edgar, R. S. 1965. *General genetics.* San Francisco: W. H. Freeman and Co.

Stern, C. 1973. *Principles of human genetics.* San Francisco: W. H. Freeman and Co.

Stern, C., and Sherwood, E. R. 1966. *The origin of genetics.* San Francisco: W. H. Freeman and Co.

Sturtevant, A. H. 1965. *A history of genetics.* New York: Harper & Row.

Watson, J. D. 1970. *Molecular biology of the gene.* New York: W. A. Benjamin.

Winchester, A. M. 1970. *Genetics.* Boston: Houghton Mifflin Co.

3 Mutation

Each gene of an organism may assume a variety of forms. A normal gene may change to another form and produce an effect on a trait different from that of the original gene. An inheritable change in the structure of the gene is known as a *mutation*. Variant traits such as albinism in humans and vestigial wings in the fruit fly are traceable to the action of altered, or mutant, genes. Mutations are the ultimate source of genetic variation. *All differences in the genes of organisms have their origin in mutation.*

Causes of Mutation

New mutations arise from time to time, and the same mutation may occur repeatedly. It is often difficult to distinguish between new mutations and old ones that occurred previously and were carried concealed in ancestors. A recessive mutant gene may remain masked by its normal dominant allele for many generations, and reveal itself for the first time only when two heterozygous carriers of the same mutant gene happen to mate.

Each gene runs the risk of changing to an alternative form. The causes of naturally occurring, or *spontaneous,* mutations are largely unknown. The environment contains a background of inescapable radiation from radioactive elements, cosmic rays, and gamma rays. It is generally conceded that the amount of background radiation is too low to account for all spontaneous mutations. In other words, only a small fraction of spontaneous mutations can be attributed to background radiation.

In 1927, the late Nobel laureate Hermann J. Muller of Indiana University announced that genes are highly susceptible to the action of X rays. By irradiating fruit flies with X rays, he demonstrated that the process of mutation is enormously speeded up. The production of mutations is dependent on the total dosage of X rays (measured in units called *roentgens*). The yield of mutations is related to the magnitude of radiation exposure. It has long been held that the mutagenic effect is the same whether the dose is given in a short time or spread over a long period. In other words, low intensities of X rays over long periods of time produce as many mutations as the same dose administered in high intensities in a short period of time. Recent experiments on mice have cast some doubt on this view, for it has been shown (at least in mice) that the mutagenic effect of a single exposure to the germ cells is greater than the effect of the same exposure administered as several smaller doses separated by intervals of time. Moreover, there does not appear to be a critical, or *threshold,* dose of roentgens below which there is no effect. In essence, no dose is so low (or "safe") as to carry no risk of inducing a mutation. Modern workers stress that any amount of radiation, no matter how little, can cause a mutation.

At the time of Muller's discovery, no one conceived that within a generation the entire population of man would be exposed to a significant increase of high-energy radiation as a consequence of the creation of the atomic bomb. The additional amount of high-energy radiation already produced by fallout from atomic explosions has undoubtedly increased the mutation rate. Most of the radiation-induced mutations are recessive and most of them are deleterious.

In discussing the hazards of X rays and other ionizing radiations, we must be careful to distinguish between *somatic damage* and *genetic damage.* Injury to the body cells of the exposed individual himself constitutes somatic damage. On the other hand, impairment of the genetic apparatus of the sex cells represents genetic damage. Typically, the genetic alterations do not manifest themselves in the individual himself but present a risk for his descendants in the immediate or succeeding generations.

There are documented records of the *somatic* consequences of exposure to the 1945 atomic blasts in Hiroshima and Nagasaki. Among 161 children born of women who were exposed to the atomic bomb while pregnant, 29 were microcephalic (head size considerably below normal) and 11 of these 29 were mentally retarded. As might have been expected, the deleterious effects were most pronounced among the infants of women who were in an early stage of pregnancy (less than 15 weeks' gestation) at the time of exposure. Moreover, most of the women who gave birth to deformed infants were less than 1.3 miles from the center (hypocenter) of the explosion. The adverse effects on the fetus diminished in frequency and severity as the distance from the hypocenter increased. Analyses have also revealed that 15 percent more people died per year in the decade 1950 to 1960 for a sample of 99,393 atomic blast survivors of all ages compared with unexposed control groups. Many of the increased deaths were due to acute leukemia, which substantiates the association found in other studies between whole-body exposure to radiation at high dose levels and the incidence of leukemia.

Estimates of the magnitude of possible *genetic* damages have been of uncertain significance. The statistical methods used are too insensitive to detect the occurrence of radiation-induced mutations. There was no greater amount of gene-determined defects to infants born to women who were pregnant at the time of the atomic blasts at Hiroshima and Nagasaki compared with populations that were not exposed. However, it must be stressed that most of the possible recessive mutations induced would not be expressed for several generations.

In an effort to measure immediate damaging genetic effects, a possible shift in the sex ratios among children of exposed parents was sought. Theoretically, if fatal, or lethal, mutations are induced on the X chromosome, they will be immediately expressed in the sons of the exposed mother (fig. 3.1). Deaths of male fetuses prior to birth can be indirectly gauged by a decrease in the number of males born to irradiated mothers. An early, limited study on the offspring of Japanese mothers who received a heavy dose of radiation revealed a slight reduction of male births, but a later study, based on a larger sample, showed no alteration of the sex ratio.

Although the actual data available have failed to reveal unequivocal genetic effects of radiation, it would be a patent fallacy to conclude that atomic radiation has had no mutagenic effect. American and British geneticists estimate that each person currently receives a total dose of 7.8 roentgens to the reproductive cells during the first 30 years of life. Of this amount, 3.1 roentgens are derived from natural background radiation, 4.6 roentgens from various medical uses of ionizing radiation, and an additional

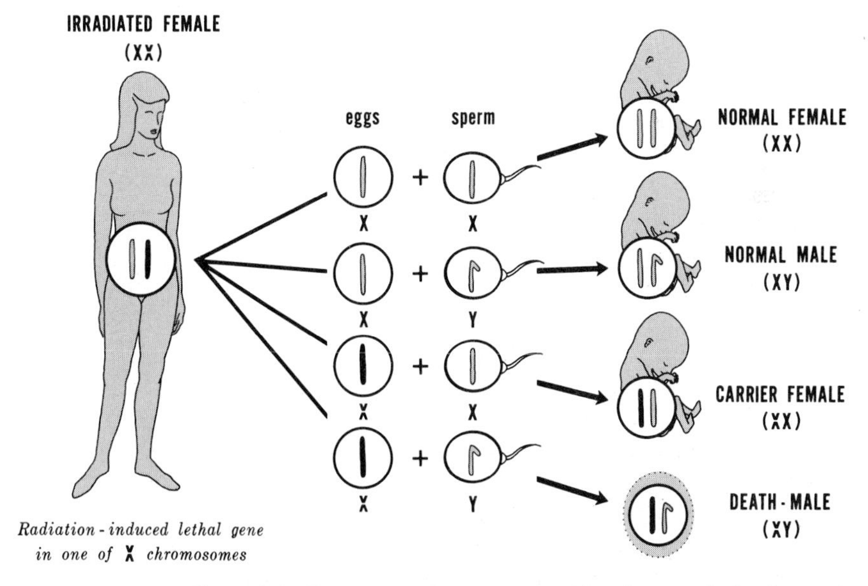

Figure 3.1 Death of male fetuses resulting from radiation-induced lethal mutations in one of the X chromosomes of the mother.

0.1 roentgen from the testing of nuclear weapons. The additional amount of radiation (0.1 roentgen) received from nuclear fallout may seem trivial. However, the exposure of our population to 0.1 roentgen is calculated to induce sufficient mutations to result in 3,750 defective offspring among 100 million births.

Chemical Nature of the Gene

During the early decades of this century, the picture of genes that had emerged was that of discrete entities, strung together along the chromosome, like individual beads on a string. This portrayal carried with it the implication that genes were not pure abstractions but actual particles of living matter. Yet, until the 1950s the chemical nature of the gene remained enigmatic. Also, no one knew precisely how genes made additional copies of themselves or how they controlled the development of a trait.

One of the finest triumphs of modern science has been the elucidation of the chemical nature of the gene. The transmission of traits from parents to offspring depends on the transfer of a specific giant molecule that carries a coded blueprint in its molecular structure. This complex molecule, the basic chemical component of the chromosome, is *deoxyribonucleic acid* (DNA). The information carried in the DNA molecule can be divided into a number of separable units, which we now recognize as the genes. Stated simply, chromosomes are primarily long strands of DNA, and genes are coded sequences of the DNA molecule. The amount of DNA present in a single unfertilized human egg has been estimated to carry information corresponding to two million genes.

The two scientists who worked together in the early 1950s to propose a configuration for the DNA molecule were Francis H. C. Crick, a biophysicist at Cambridge University, and James D. Watson, an American student of virology who was then studying chemistry at Cambridge on a postdoctoral fellowship. With the invaluable aid of X ray pictures of DNA crystals prepared by Maurice H. F. Wilkins, a biophysicist at King's College in London, Watson and Crick built an inspired model in metal of DNA's configuration. This achievement won Watson, Crick, and Wilkins the coveted Nobel Prize for physiology and medicine in 1962.

The remarkable feature of DNA is its simplicity of design. The double-stranded DNA molecule is shaped like a twisted ladder (fig. 3.2). The two parallel strands of the ladder are twisted around each other somewhat like the supporting frameworks of a spiral staircase. The twisted supports of the ladder are made up of alternating units of sugar (deoxyribose) and phosphate molecules, while the cross-links or rungs are composed of specific nitrogen-containing ring compounds, or nitrogenous bases. There are two classes of nitrogenous bases, the larger, two-ring purines and the smaller, one-ring pyrimidines. Each rung of the ladder consists of one purine

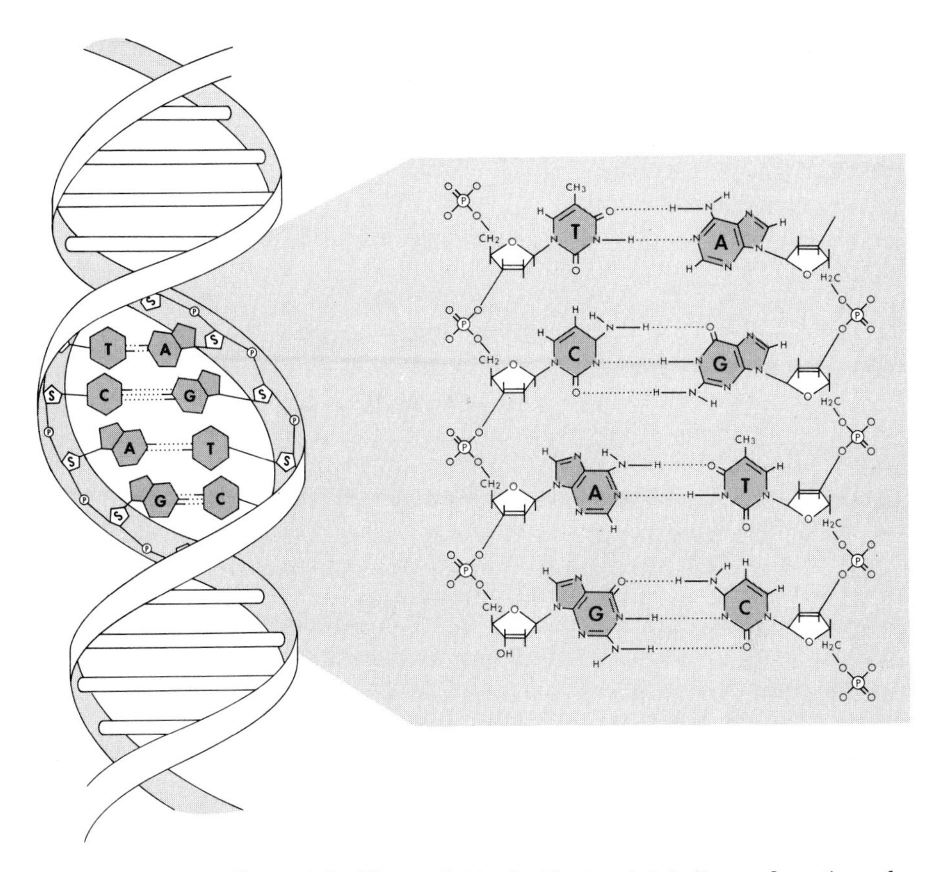

Figure 3.2 Watson-Crick double-stranded helix configuration of deoxyribonucleic acid (DNA). The backbone of each twisted strand consists of alternating sugar (S) and phosphate (P) residues. Enlarged view on right shows that the larger, two-ring purines (adenine or guanine) lay opposite the smaller, one-ring pyrimidines (cytosine or thymine). The nitrogenous bases are held together by hydrogen bonds, three between cytosine (C) and guanine (G), and two between adenine (A) and thymine (T).

coupled to one pyrimidine. The arrangement of the bases is not haphazard: adenine (A) in one chain is normally joined with thymine (T) in the other chain, and guanine (G) is typically linked with cytosine (C). Along any one chain, any sequence of the bases is possible, but if the sequence along one chain is given, then the sequence along the other is automatically determined, because of the precise pairing rule (A = T and G = C). The combination of one purine and one pyrimidine to make up each cross connection is conveniently called a *base pair*.

The DNA molecule has a structure that is sufficiently versatile to account for the great variety of different genes. The four bases (A, C, G, and

T) may be thought of as a four-letter alphabet or code. A given gene owes its unique character to a specific order of the bases, just as words in our language differ according to the sequence of letters of the alphabet or as a telegraphic message becomes comprehensible by the varied combinations of dots and dashes. The number of ways in which the nitrogenous bases can be arranged in the DNA molecule is exceedingly large. Current estimates indicate that a single gene is a linear sequence of approximately 1,500 base pairs.

Self-Copying of DNA Molecule

One fundamental property that has long been ascribed to the gene is its ability to make an exact copy of itself. If the gene is really a linear sequence of base pairs, then the Watson-Crick model of DNA must be able to account for the precise reproduction of the sequences of the bases. The exact replication of the double helix can be visualized with little difficulty. The DNA molecule consists of two parts, each of which is the complement of the other. Accordingly, each single chain can serve as a template to guide the formation of a new companion chain (fig. 3.3). The two parallel chains separate, breaking the hydrogen bonds that hold together the paired bases. Each chain then attracts new base units from among the supply of free units always present in the cell. As separation takes place, each separated lengthwise portion of the chain can begin to form a portion of the new chain. Eventually, the original double helix has produced two exact replicas of itself. If the original two chains are designated *A* and *B*, then *A* will direct the formation of a new *B* and the old *B* will guide the production of a new *A*. Where one *AB* molecule existed previously, two *AB* molecules, exactly like the original, exist afterward.

The mechanism of DNA self-duplication (or replication) is in accord with our understanding of the duplication of chromosomes during the division of a cell. When the fertilized egg divides into many daughter cells, the chromosomes regularly produce copies of themselves with each cell division. As a result, each of the millions of cells of the developing embryo carries faithful copies of the chromosomes of the fertilized egg. The process by which new cells in the organism arise from preexisting cells is termed *mitosis*. The precise similarity of the new, or daughter, chromosomes to the original, or parent, chromosome is ensured by the mode of replication of the DNA molecule. When two identical molecules of DNA have been formed, one copy is passed on to each of the two daughter cells during mitosis. Thus, each daughter cell receives the same set of genetic instructions as was originally present in the parent cell.

Molecular Mechanism of Mutation

DNA replication is subject to error. A mutation can be envisioned as a chance mishap to one of the paired bases in the DNA molecule. A single sub-

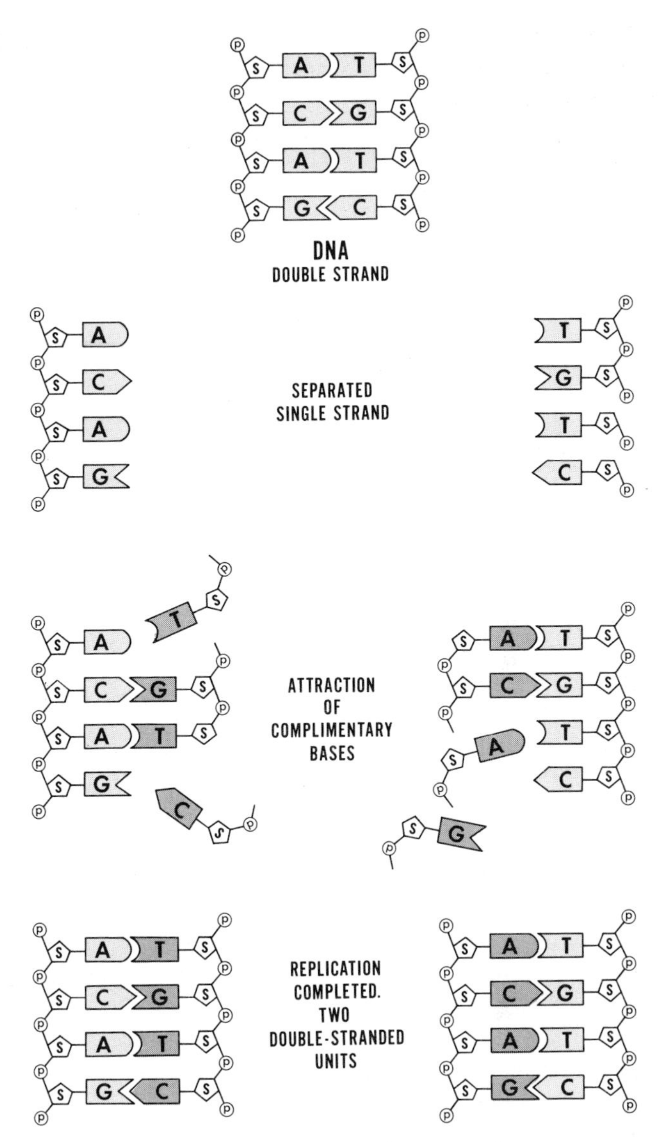

Figure 3.3 Replication of DNA. The parental strands separate, each serving as a guide or template for the synthesis of a new daughter strand. Each new DNA molecule contains one parental and one daughter strand.

stitution of a C–G pair for a T–A may be sufficient to alter the character of the gene. One possible way in which such a substitution might occur is shown in figure 3.4. It has been established that each of the bases in DNA can exist in several rare alternative forms known as *tautomers*. The tautomeric alternatives differ in the positions at which certain atoms are attached in the molecule. The significance of the tautomeric alternatives lies in the changed

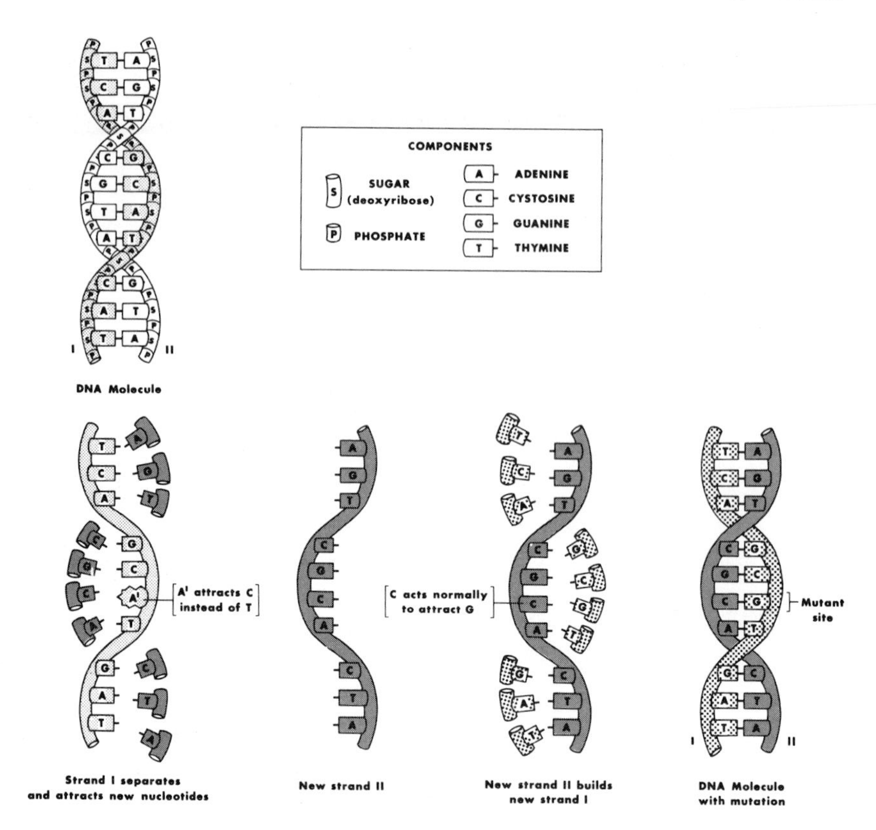

Figure 3.4 Mutation at the molecular level. When cell division occurs, the two twisted strands separate and each strand attracts unbound nucleotides (containing nitrogenous bases) to rebuild the DNA molecule. This particular illustration shows events through two successive cell divisions, commencing with strand I in the original DNA molecule. During the first division, an adenine radical (A) undergoes a chemical change to A', which misattracts cytosine (C) instead of thymine (T). At the next division, cytosine (C) attracts its customary partner, guanine (G). The net result is that the granddaughter DNA molecule contains a C-G pair where a T-A pair was formerly located. This highly localized change in one of the pairs of nitrogenous bases may qualify as a gene mutation.

pairing qualities they impart. Whereas, for example, the normal form of adenine pairs with thymine, one of its rare tautomeric forms actually attracts and couples with cytosine. Such a tautomeric shift in the adenine molecule may thus make possible a new purine-pyrimidine base pair arrangement in the DNA molecule. As seen in figure 3.4, an adenine (labeled A^1) has undergone a tautomeric shift and attracts the wrong partner (cytosine instead of thymine). At the next replication of the strand, the misattracted cytosine acts normally to join with guanine. Hence, a C–G pair is established where a T–A pair was formerly located. At this point in the DNA molecule, the gene is modified and may be expected to produce a mutant effect.

Many mutations are simply base-pair switches in the DNA molecule. Changes at the molecular level within the gene have been called *point* (or *gene*) *mutations*. Grosser alterations involving the chromosomes, such as the loss of a segment of a chromosome, have been termed *chromosomal aberrations*.

The mutation process is generally thought to be an uncontrollable chance phenomenon. Yet, the phenomenal surge of knowledge today promises to provide unprecedented opportunities in the future for man to manipulate the DNA molecule in organisms, including himself. The DNA molecule has already been synthesized artificially in the laboratory. The test tube synthesis of a faithful copy of the DNA of a virus was achieved in 1967 by Nobel laureate Arthur Kornberg of Stanford University. Then, in 1969, a team of Harvard researchers, led by Jonathan Beckwith, succeeded in isolating a specific gene of a bacterial cell. In 1970, the biochemist Har Gobind Khorana and his colleagues, then at the University of Wisconsin, painstakingly assembled a single gene from relatively simple laboratory chemicals containing carbon, hydrogen, oxygen, and phosphorus. The gene specifies the production of a large molecule in the cytoplasm of yeast cells. The artificially synthesized gene is relatively simple, containing only 77 pairs of bases. In 1976, Khorana, now at the Massachusetts Institute of Technology, reported the synthesis of an artificial gene that is capable of functioning in a living cell. The synthetic gene, when inserted in a mutant strain of a bacteriophage, was able to replace the defective gene in that strain. These remarkable feats bring closer the day that a hereditary defect in man may be averted by incorporating a synthetic normal gene into a defective human gene complex, or by artificially inducing the abnormal gene to "back-mutate" to its normal form.

Language of Life

It is now widely accepted that genes provide the information to direct the synthesis of proteins in the cell. The gene—a linear sequence of bases in the DNA—codes for the precise sequence of amino acids that compose a protein. The properties of any protein are determined by the number, identity, and arrangement of the amino acid residues (components). In a large sense, the molecular basis of the genotype is DNA; the molecular basis of the phenotype is protein. Ultimately all inherited differences in individuals are expressed in their bodily structure through changes in specific proteins.

Amino acids, the basic building blocks of proteins, are arranged in long chains, called polypeptides. Each polypeptide chain may be several hundred amino acid units in length. The number of possible arrangements of the different amino acids in a given protein is unbelievably large. In a polypeptide chain made up of 500 amino acid units, the number of possible patterns (given 20 different amino acids) can be expressed by the number 1 followed by 1,100 zeroes. How, then, does a cell form the particular amino acid patterns it requires out of the colossal number of patterns that are possible?

We may presume that the sequence of bases of the DNA molecule is in some way the master pattern, or code, for the sequence of the amino acids in the polypeptide chains of the cellular proteins. In 1954, the British physicist George Gamow suggested that each amino acid is dictated by one sequence of three bases in the DNA molecule. As an example, the sequence cytosine-thymine-thymine (CTT) in the DNA molecule might designate that the amino acid glutamic acid is to be incorporated in the formation of a protein molecule, such as hemoglobin (fig. 3.5). Thus, the DNA code is to be found in triplets—that is, three bases taken together code one amino acid. It should be noted that only one of the two strands of the DNA molecule serves as the genetic code. Biochemical chaos would result if both complementary strands of DNA were to encode information.

The importance of the genetic code cannot be overstated. Let us assume the following sequence of bases in one of the strands of the DNA molecule: . . . CGT ATC GTA AGC . . . , and that the triplets in this sequence specify four amino acids, designated R, I, P, and E. In other words, the following code exists: CGT = R, ATC = I, GTA = P, and AGC = E. This section of the DNA molecule thus specifies RIPE, and the message to make RIPE is passed on by this particular sequence from the nucleus to the cytoplasm of the cell. The message continues to flow out in the living cell, and RIPE will be made, copy after copy.

But what if a chance mishap occurs in one of the triplets? Let us say that T in the second triplet is substituted for A, so that the triplet reads TTC instead of ATC, which would specify A instead of I. The word would now be RAPE instead of RIPE. In a publishing analogy, think of the difference that this continually misprinted word would have throughout a novel. And just as a misprinted word can alter or destroy the meaning of a sentence, so can an altered protein in the body fail to express its intended purpose. Sometimes the error is not tragic, but often the organism is debilitated by the misprint.

As a striking example, we may consider the case of individuals afflicted with sickle-cell anemia. The hemoglobin molecule in sickle-cell anemic patients is biochemically abnormal. In 1956, Vernon Ingram, then at Cambridge University, succeeded in breaking down hemoglobin, a large protein molecule, into several peptide fragments containing short sequences of identifiable amino acids. Normal hemoglobin and sickle-cell hemoglobin yielded the same array of peptide fragments, with a single exception. In one of the peptide fragments of sickle-cell hemoglobin, the amino acid *glutamic acid* had been replaced at one point in the chain by *valine* (fig. 3.5). The sole difference in chemical composition between normal and sickle-cell

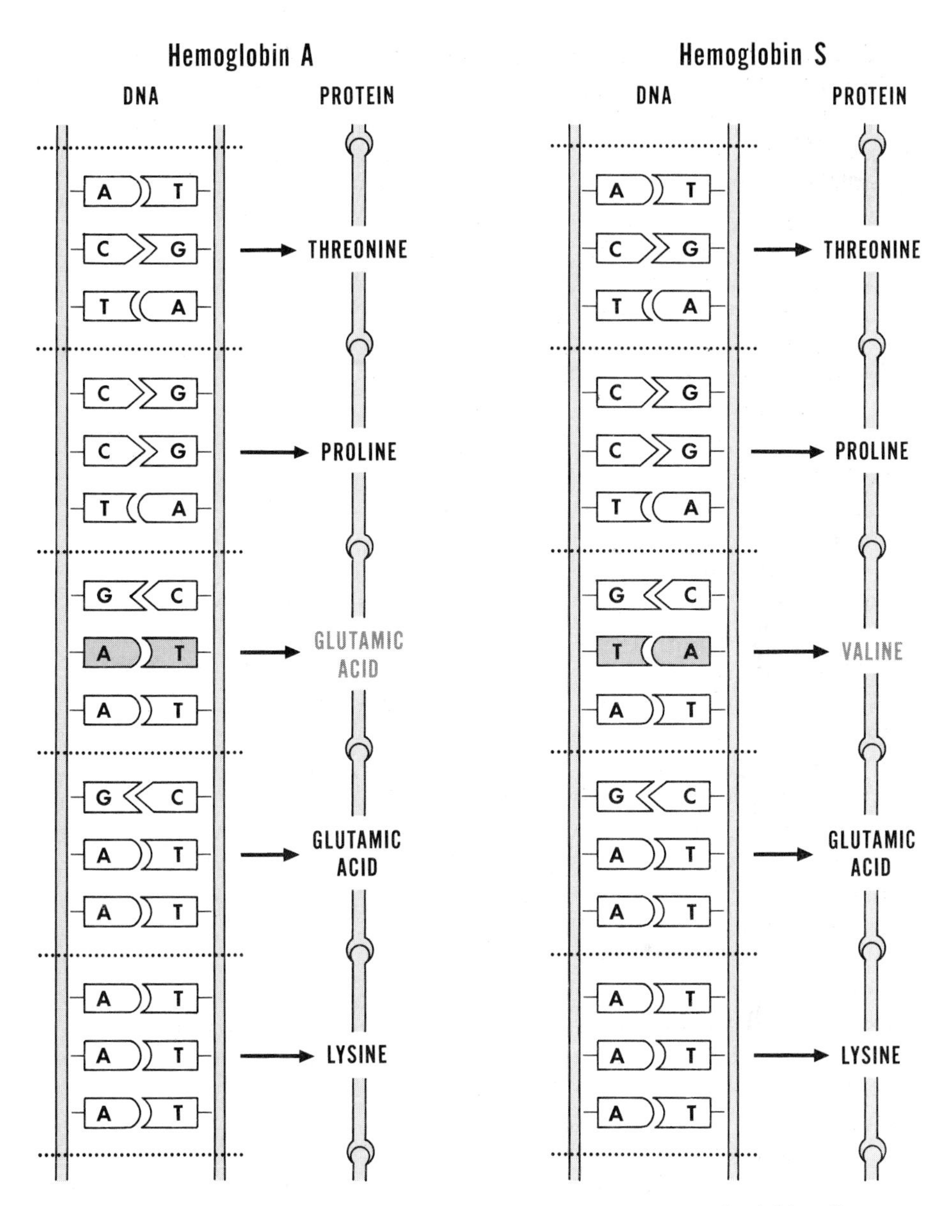

Figure 3.5 The abnormal hemoglobin that occurs in sickle-cell anemia (hemoglobin S) is the consequence of an alteration of a single base of the triplet of DNA that specifies a particular amino acid (glutamic acid) in normal hemoglobin (hemoglobin A). A simple base-pair switch, or mutation, in the DNA molecule results in the replacement of glutamic acid in hemoglobin A by another amino acid, valine, in hemoglobin S. (Based on studies by Vernon M. Ingram.)

hemoglobin is the substitution of only one amino acid unit among several hundred. The detrimental effect of sickle-cell anemia is thus traceable to an exceedingly slight alteration in the structure of the protein molecule. This, in turn, is associated with a highly localized genetic change, or mutation, in one of the base pairs in the DNA molecule in the chromosome (fig. 3.5).

The DNA molecule, like a tape recording, carries specific messages for the synthesis of a wide variety of proteins. However, there are complexities in the process of protein synthesis. For example, the DNA molecule does not directly form protein but works in a complex way through a secondary form of nucleic acid, *ribonucleic acid,* or *RNA*. It is sufficient for our present purpose to recognize that a gene is a coded sequence in the DNA molecule. From a functional point of view, we may say that the gene is *a section of the DNA molecule (about 1,500 base pairs) involved in the determination of the amino acid sequence of a single polypeptide chain of a protein.*

Frequency of Mutations

The rate at which any single gene mutates is generally low, but constant. The average rate of mutation per gene in the fruit fly (*Drosophila melanogaster*) is thought to be about 1 in 100,000 gametes. In other words, any given gene, on the average, mutates approximately once in every 100,000 sperm cells or egg cells produced. Some genes change more often than others. Some indication of the rate of mutation can be obtained from the incidence of a genetic disorder in the population. In humans, a form of dwarfism known as *achondroplasia* occurs in 1 child in nearly 12,000 born to normal parents. Affected infants are small and disproportionate, with abnormally short arms and legs (fig. 3.6). The defect is present at birth, and most achondroplastic dwarfs are stillborn or die in infancy. This well-known abnormality is caused by a dominant mutant gene. Since the parents are normal, each new case of this dominant disorder must result from a newly mutated gene originating in either the sperm or the egg cell. Twelve thousand births represent a total contribution of 24,000 gametes. Accordingly, a new mutation for this peculiar type of dwarfism arises at a rate of 1 in 24,000 gametes, or roughly 4 in 100,000 gametes (4×10^{-5}).

We hasten to point out that the foregoing calculated rate of mutation is undoubtedly too high for a single gene. At the outset, the estimate is unreliable because it is not based on a sufficiently large population sample of the incidence of achondroplasia. In addition, there is evidence that this deformity may result from mutations at two or more different gene loci. Hence, the rate of 4×10^{-5} may actually be the product of two or three separate rates. Other methods of gauging mutation rates also have noteworthy complications. In our present state of knowledge, the mutation rate in man for any single gene locus must be considered as a very rough estimate. The consensus is that the mutation rate in man is in the order of magnitude of one per 100,000 gametes (1×10^{-5}).

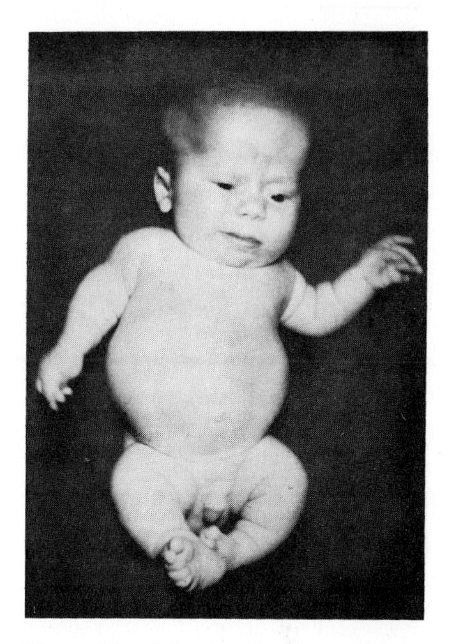

Figure 3.6 Achondroplastic dwarfism, a dominant genetic disorder in which the affected infant has inherited abnormally short arms and legs. (Courtesy of Dr. Norman Woody.)

Overall Mutation Rate

Although the mutation rate of a single gene may be low, the mutability of the organism as a whole is obviously much higher when we account for the total complement of genes possessed by an individual. A single unfertilized human egg contains an estimated three million (3×10^6) base pairs in its nuclear DNA. This incredible number of nucleotide pairs can carry information corresponding to two million genes. This number of genes is overwhelming when one considers the comparatively small number of genetic defects in man with a simple mode of inheritance. We may presume that not all single base changes (mutations) have adverse effects.

Let us suppose that there are 50,000 genes in man in which a mutation (simple base substitution) can lead to a phenotypically noticeable deviation. The average rate of mutation per gene in man is generally held to be about 1 per 100,000 gametes. If the value of 1/100,000 mutation per gene (1×10^{-5}) is multiplied by the conservative figure of 50,000 genes (5×10^{-4}), then the average number of new mutations per gamete is 0.50. Thus, one of every two gametes produced by a person would bear a new mutant gene capable of having an adverse effect.

Each offspring, of course, is the product of two gametes. Let us envision the random union of sperm cells and egg cells to form zygotes, and say that

50 percent of all gametes bear one newly mutated gene. The chance that an egg cell containing a newly mutated gene will be fertilized by a sperm cell carrying a new mutated gene is 1 in 4, or 25 percent. (Recall that the probability that both coins will show heads when tossed together is $1/2 \times 1/2 = 1/4$.) The chance that a zygote will be free of a new mutant gene is also 25 percent (like the probability of two tails). Finally, the chance that a zygote will carry one newly mutated gene contributed by either the sperm cell or the egg cell is 50 percent (similar to the probability of the combination of a head and a tail when two coins are tossed together). It should be apparent that 75 percent of all human offspring would contain at least one newly mutated gene. This assertion, based on crude computations, is in accord with currently accepted notions. As expressed by the English geneticist Harry Harris, every newborn infant, on the average, may be "expected, as a result of a new mutation in either of its parents, to synthesize at least one structurally variant enzyme or protein." Viewed in this manner, the phenomenon of mutation should become very real to us.

It is of interest that all of human pregnancies that continue longer than four or five weeks, 12 to 15 percent end as abortions before the end of the 27th week of pregnancy, and 2 percent terminate as stillbirths. Moreover, the proportion of zygotes lost earlier than the fourth week of pregnancy has been estimated at 10 to 15 percent. Taken all together, 30 percent of all human embryos fail to survive *in utero*. Not all failures, of course, are due to detrimental mutations, but the mutation phenomenon undoubtedly is an important contributing factor.

Harmful Nature of Mutations

Most by far of the gene mutations observable today in organisms are changes for the worse. This is not unexpected. Existing populations of organisms are products of a long evolutionary past. The genes that are now normal to the members of a population represent the most favorable mutations selectively accumulated over eons of time. The chance that a new mutant gene will be more advantageous than an already established favorable gene is slim. Nonetheless, if the environment were to change, the previously adverse mutant gene might prove to be beneficial in the new environmental situation. The microscopic water flea, *Daphnia,* thrives at a temperature of 20°C and cannot survive when the temperature rises to 27°C. A mutant strain of this water flea is known that requires temperatures between 25°C and 30°C and cannot live at 20°C. Thus, at high temperatures, the mutant gene is essential to the survival of the water fleas. This little episode reveals an important point: *A mutation that is inferior in the environment in which it arose may be superior in another environment.*

The process of mutation furnishes the genetic variants that are the raw materials of evolution. Ideally, mutations should arise only when advantageous, and only when needed. This, of course, is fanciful thinking. Mutations occur irrespective of their usefulness or uselessness. The mutations responsible for achondroplasia and retinoblastoma (malignant eye tumors) in man are certainly not beneficial. But novel inheritable characters repeatedly arise as a consequence of mutation. Only one mutation in several thousands might be advantageous, but this one mutation might be important, if not necessary, to the continued success of a population. The harsh price of evolutionary potentialities for a population is the continual occurrence and elimination of mutant genes with detrimental effects. In evolutionary terms, a population must depend on the occasional errors that occur in the copying process of its genetic material if it is to continue to evolve.

Selected Readings

Asimov, I. 1962. *The genetic code.* New York: Signet Science Library.

Crick, F. H. C. 1954. The structure of the hereditary material. *Scientific American,* October, pp. 54–61.

Crick, F. H. C. 1962. The genetic code. *Scientific American,* October, pp. 66–74.

Crick, F. H. C. 1966. The genetic code: III. *Scientific American,* October, pp. 55–62.

Crow, J. F. 1959. Ionizing radiation and evolution. *Scientific American,* September, pp. 138–60.

Dobzhansky, T. 1964. *Heredity and the nature of man.* New York: The New American Library.

Ingram, V. 1958. How do genes act? *Scientific American,* January, pp. 68–74.

Kornberg, A. 1968. The synthesis of DNA. *Scientific American,* October, pp. 64–78.

Miller, R. W. 1969. Delayed radiation effects in atomic-bomb survivors. *Science* 166: 569–74.

Muller, H. J. 1955. Radiation and human mutation. *Scientific American,* November, pp. 58–68.

Nirenberg, M. 1963. The genetic code: II. *Scientific American,* March, pp. 80–94.

Papazian, H. P. 1967. *Modern genetics.* New York: W. W. Norton & Co.

Schull, W. J., and Neel, J. V. 1958. Radiation and the sex ratio in man. *Science* 128: 343–48.

Sinsheimer, R. 1962. Single-stranded DNA. *Scientific American,* July, pp. 109–16.

United Nations Scientific Committee. 1967. *The genetic risk from radiation.* New York: United Nations.

Wallace, B. 1966. *Giant molecules, and evolution.* New York: W. W. Norton & Co.

Wallace, B., and Dobzhansky, T. 1959. *Radiation, genes, and man.* New York: Holt, Rinehart and Winston.

Watson, J. D. 1968. *The double helix.* New York: Atheneum Publishers.

4

Genetic Equilibrium

The opening chapter introduced us to a natural population of bullfrogs that conspicuously contained at one time several hundred multilegged variants. We surmised that the multilegged anomaly was an inherited condition, transmitted by a detrimental recessive gene. The multilegged frogs disappeared in nature as dramatically as they appeared. They unquestionably failed to reproduce and leave descendants. Now, let us imagine that the multilegged frogs were as reproductively fit as their normal kin. Would the multilegged trait eventually still be eliminated from the population?

A comparable question was posed to the English geneticist R. C. Punnett at the turn of the century. He was asked to explain the prevalence of blue eyes in man in view of the acknowledged fact that the blue-eyed condition was a recessive characteristic. Would it not be the case that the dominant brown-eyes trait would in time supplant the blue-eyed state in the human population? The answer was not self-evident, and Punnett sought out his colleague Godfrey H. Hardy, the astute mathematician at Cambridge University. Hardy had only a passing interest in genetics, but the problem intrigued him as a mathematical one. The solution, which we shall consider below, ranks as one of the most fundamental laws of genetics and evolution. As fate has it, Hardy's formula was arrived at independently in the same year (1908) by a physician, Wilhelm Weinberg, and the well-known equation presently bears both their names.

Mendelian Inheritance

We may recall that the genetic constitutions, or genotypes, of the normal and multilegged frogs have been designated as *AA* (normal), *Aa* (normal but a carrier), and *aa* (multilegged). The kinds and proportions of offspring that can arise from matings involving the three genotypes are illustrated in figure

4.1. Six different types of matings are possible. The mating *AA* × *AA* gives rise solely to normal homozygous offspring, *AA*. Two kinds of progeny, *AA* and *Aa* in equal proportions, result from the cross of a homozygous normal parent (*AA*) and a heterozygous parent (*Aa*). The mating of two heterozygotes (*Aa* × *Aa*) produces *AA, Aa,* and *aa* offspring in the classical Mendelian ratio of 1:2:1. Only heterozygous offspring (*Aa*) emerge from the mating *AA* × *aa*. Both heterozygous (*Aa*) and recessive (*aa*) progeny, in equal numbers, arise from the cross of a heterozygous parent (*Aa*) and a recessive parent (*aa*). Lastly, two recessive parents (*aa* × *aa*) produce only recessive offspring (*aa*).

These principles of Mendelian inheritance merely inform us that certain kinds of offspring can be expected from certain types of matings. If we are interested in following the course of a population from one generation to the next, then additional factors enter the scene.

Random Mating

One important factor that influences the genetic composition of a population is the system of mating among individuals. The simplest scheme of breeding activity in a population is called *random mating,* wherein any one individual has an equal chance of pairing with any other individual. Random mating does not mean promiscuity; it simply means that those who choose each other as mating partners do not do so on the basis of similarity or dissimilarity in a given trait or gene.

The absence of preferential mating in a population has interesting consequences. Let us suppose that random mating prevails in our particular population of bullfrogs. This assemblage of frogs is ordinarily very large, numbering several thousand individuals. For ease of presentation, however, the size of the population is reduced to 48 males and 48 females. Moreover, for each sex, we may simplify the mathematical computations by assuming that 36 are normal (12 *AA* and 24 *Aa*) and 12 are multilegged (*aa*). Accordingly, one-quarter of the individuals of each sex are homozygous dominant, one-half are heterozygous, and one-quarter are recessive. Now, if mating occurs at random, will the incidence of multilegged frogs decrease, increase, or remain the same in the next generation?

The problem may be approached by determining how often a given type of mating occurs. Here we will use the multiplication rule of probability: *The chance that two independent events will occur together is the product of their chances of occurring separately*. The proportion of *AA* males in our arbitrary bullfrog population is 1/4. We may also say that the chance that a male bullfrog is *AA* is 1/4. Likewise the probability that a female bullfrog is *AA* is 1/4. Consequently, the chance that an *AA* male will "occur together," or mate, with an *AA* female is 1/16 (1/4 × 1/4). The computations for all

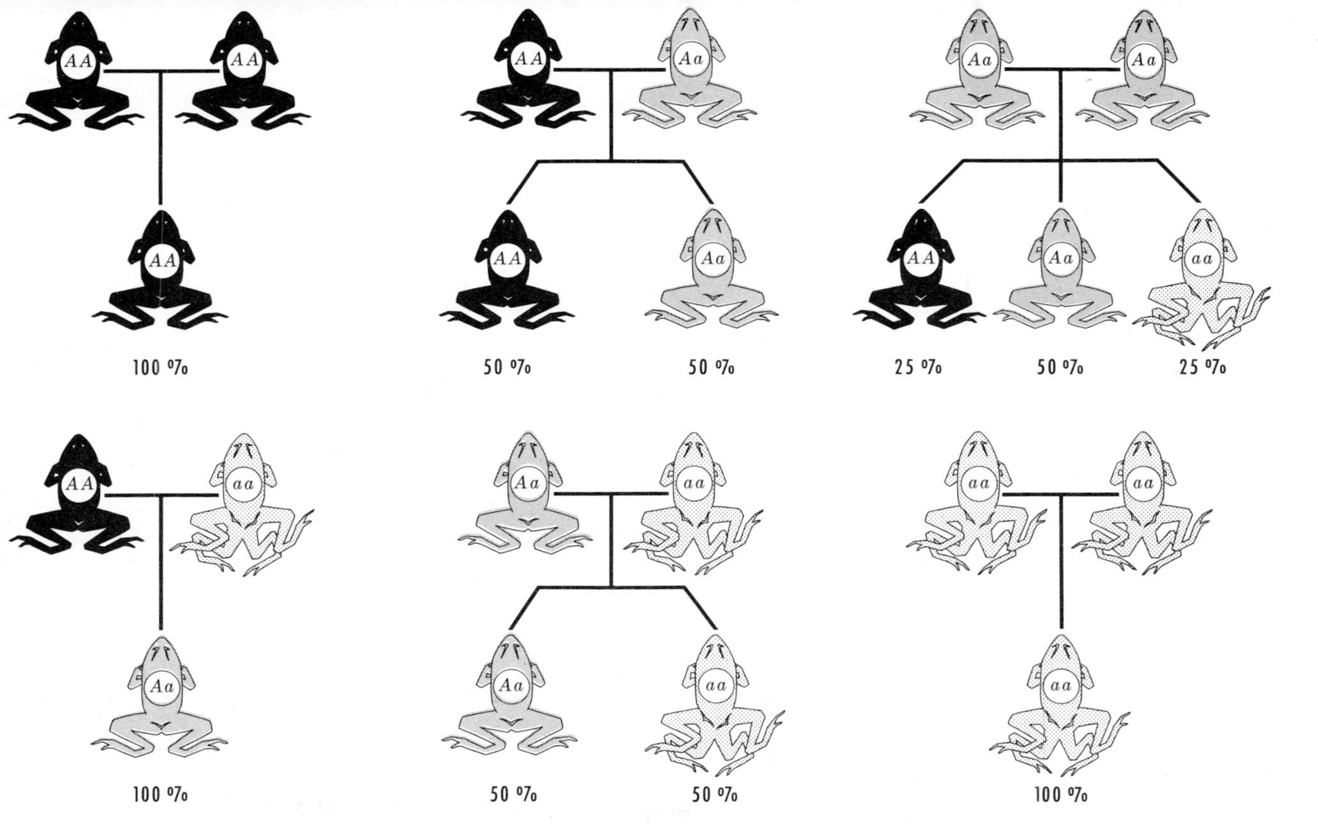

Figure 4.1 Six possible mating types with respect to one pair of genes, and the kinds and percentages of offspring from each type of mating. Normal frogs are either homozygous dominant (AA) or heterozygous (Aa); multilegged frogs are recessive (aa). The sex of the parent is not denoted; in crosses of unlike genotypes (such as AA x Aa), either genotype may be the male or the female.

types of matings can be facilitated by coupling the males and females in a multiplication table, as shown in table 4.1.

Table 4.1
Random Mating

Female (♀)	Male (♂)		
	1/4 *AA*	2/4 *Aa*	1/4 *aa*
1/4 *AA*	1/16 *AA* x *AA*	2/16 *AA* x *Aa*	1/16 *AA* x *aa*
2/4 *Aa*	2/16 *Aa* x *AA*	4/16 *Aa* x *Aa*	2/16 *Aa* x *aa*
1/4 *aa*	1/16 *aa* x *AA*	2/16 *aa* x *Aa*	1/16 *aa* x *aa*

Table 4.1 shows that there are nine combinations of mated pairs, and that some types occur more frequently than others. It may be helpful to express the frequencies in terms of actual numbers. Thus, for a total of 48 matings, 3 ($= 1/16 \times 48$) would be *AA* ♀ × *AA* ♂, 6 ($= 2/16 \times 48$) would be *AA* ♀ × *Aa* ♂, 12 ($= 4/16 \times 48$) would be *Aa* ♀ × *Aa* ♂, and so forth. The numbers of each type of mating are listed in table 4.2.

Table 4.2
First Generation of Offspring

Type of Mating (Female x Male)	Number of Each Type of Mating*	Number of Offspring		
		AA	*Aa*	*aa*
AA x *AA*	3	12		
AA x *Aa*	6	12	12	
AA x *aa*	3		12	
Aa x *AA*	6	12	12	
Aa x *Aa*	12	12	24	12
Aa x *aa*	6		12	12
aa x *AA*	3		12	
aa x *Aa*	6		12	12
aa x *aa*	3			12
		48 (25%)	96 (50%)	48 (25%)

* Based on a total of 48 matings.

Our next step is to ascertain the kinds and proportions of offspring from each mating. We shall assume that each mated pair yields the same number of offspring—for simplicity, four offspring. (This is an inordinately small number, as a single female bullfrog can deposit well over 10,000 eggs.) We also take for granted that the genotypes of the four progeny from each mating are those that are theoretically possible in Mendelian inheritance (see fig. 4.1). For example, if the parents are *Aa* × *Aa,* their offspring will be 1 *AA*, 2 *Aa,* and 1 *aa.* In another instance, if the parents are *AA* × *Aa,* then the offspring will be 2 *AA* and 2 *Aa*. The outcome of all crosses is shown in table 4.2. It is important to note that the actual numbers of offspring recorded in table 4.2 are related to the frequencies of the different types of matings. For example, the mating of an *AA* female with an *Aa* male occurs six times; hence, the numbers of offspring are increased sixfold (from 2 each of *AA* and *Aa* to 12 each of the two genotypes).

An examination of table 4.2 reveals that the kinds and proportions of individuals in the new generation of offspring are exactly the same as in the parental generation. There has been no change in the ratio of normal frogs (75 percent *AA* and *Aa*) to multilegged frogs (25 percent *aa*). In fact, the proportions of phenotypes (and genotypes) will remain the same in all successive generations, provided that the system of random mating is continued.

Gene Frequencies

There is a less tedious method of arriving at the same conclusion. Rather than figure out all the matings that can possibly occur, we need only to consider the genes that are transmitted by the eggs and sperm of the parents. Let us assume that each parent produces only 10 gametes. The 12 homozygous dominant males (*AA*) of our arbitrary initial population can contribute 120 sperm cells to the next generation, each sperm containing one *A*. The 24 heterozygous males (*Aa*) can transmit 240 gametes, 120 of them with *A* and 120 with *a*. The remaining 12 recessive males (*aa*) can furnish 120 gametes, each with *a*. The total pool of genes provided by all males will be 240 *A* and 240 *a,* or 50 percent of each kind. Expressed as a decimal fraction, the frequency of gene *A* is 0.5; of *a,* 0.5.

Since the females in our population have the same genetic constitutions as the males, their gametic contribution to the next generation will also be 0.5 *A* and 0.5 *a*. The eggs and sperm can now be united at random in a genetical checkerboard (fig. 4.2).

It should be evident from figure 4.2 that the distribution of genotypes among the offspring is 0.25 *AA* : 0.50 *Aa* : 0.25 *aa*. The random union of eggs and sperm yields the same result as the random mating of parents (refer to table 4.2). Thus, using two different approaches, we have answered the question posed in the introductory remarks to this chapter. *If the multilegged frogs are equally as fertile as the normal frogs and leave equal numbers of offspring each generation, then these anomalous frogs will persist in the population with the same frequency from one generation to the next.*

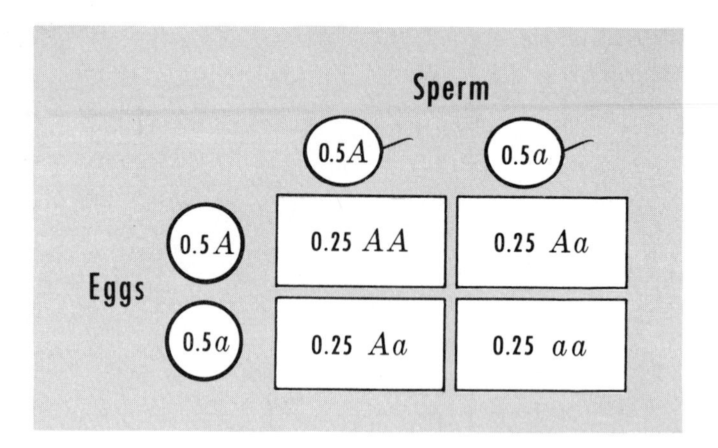

Figure 4.2 Random union of eggs and sperm yields the same outcome as the random mating of parents (refer to table 4.2).

Hardy-Weinberg Law

A population in which the proportions of genotypes remain unchanged from generation to generation is in *equilibrium*. The fact that a system of random mating leads to a condition of equilibrium was uncovered independently by G. H. Hardy and W. Weinberg, and has come to be widely known as the *Hardy-Weinberg Law*. This law states that the proportions of *AA*, *Aa*, and *aa* genotypes, as well as the proportions of *A* and *a* genes, will remain constant from generation to generation provided that the bearers of the three genotypes have equal opportunities of producing offspring in a large, randomly mating population.

The above statement can be translated into a simple mathematical expression. If we let p be the frequency of the gene *A* in the population, and q equal the frequency of its allele, *a*, then the distribution of the genotypes in the next generation will be p^2 *AA* : $2pq$ *Aa* : q^2 *aa*. This relationship may be verified by the use, once again, of a genetical checkerboard (fig. 4.3). Mathematically inclined readers will recognize that $p^2 : 2pq : q^2$ is the algebraic expansion of the binomial $(p + q)^2$. The frequencies of the three genotypes (0.25 *AA* : 0.50 *Aa* : 0.25 *aa*) in our bullfrog population under the system of random mating is the expanded binomial $(0.5 + 0.5)^2$.

We may consider another arbitrary population in the equilibrium state. Suppose that the population consists of 16 *AA*, 48 *Aa*, and 36 *aa* individuals. We may assume, as before, that 10 gametes are contributed by each individual to the next generation. All the gametes transmitted by the 16 *AA* parents (numerically, 160) will contain the *A* gene, and half the gametes (240) provided by the *Aa* parents will bear the *A* gene. Thus, of the 1,000 total gametes in the population, 400 will carry the *A* gene. Accordingly, the

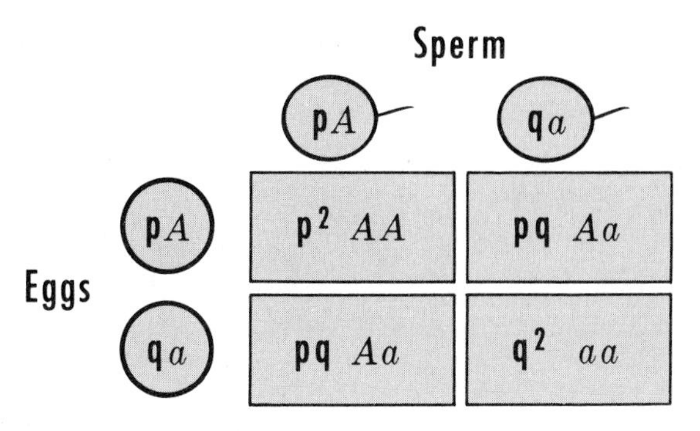

Figure 4.3 The distribution of genotypes in the next generation is p^2 *AA* : $2pq$ *Aa* : q^2 *aa* (Hardy-Weinberg formula).

frequency of gene *A* is 0.4 (designated p). In like manner, it can be shown that the frequency of gene *a* is 0.6 (q). Substituting the numerical values for p and q in the Hardy-Weinberg formula, we have:

p^2 *AA* : $2pq$ *Aa* : q^2 *aa*
$(0.4)^2$ *AA* : $2(0.4)\ (0.6)$ *Aa* : $(0.6)^2$ *aa*
0.16 *AA* : 0.48 *Aa* : 0.36 *aa*.

Hence, the proportions of the three genotypes are the same as those of the preceding generation.

Implications

The Hardy-Weinberg Law is entirely theoretical. The set of underlying assumptions can scarcely be fulfilled in any natural population. We implicitly assume the absence of recurring mutations, the absence of any degree of preferential matings, the absence of differential mortality or fertility, the absence of immigration or emigration of individuals, and the absence of fluctuations in gene frequencies due to sheer chance. But therein lies the significance of the Hardy-Weinberg law. In revealing the conditions under which evolutionary change cannot occur, it brings to light the possible forces that could cause a change in the genetic composition of a population. The Hardy-Weinberg Law thus depicts a static situation. There are several factors or forces that profoundly modify the gene frequencies in natural populations. In the next chapter, we shall consider how one of the forces, natural selection, or differential reproduction, can lead to dynamic evolutionary changes.

It should also be clear that a recessive trait, such as blue eyes in man, will not become rare just because it is governed by a recessive gene. Nor can the dominant brown-eyed condition become widespread simply by virtue of its dominance. Whether a given gene is common or rare is controlled by other factors, particularly natural selection.

Estimating the Frequency of Heterozygotes

It comes as a surprise to many readers to discover that the heterozygotes of a rare recessive abnormality are rather common instead of being comparatively rare. Recessive albinism may be used as an illustration. The frequency of albinos is about 1/20,000 in human populations. When the frequency of the homozygous recessive (q^2) is known, the frequency of the recessive gene (q) can be calculated, as follows:

$$q^2 = 1/20,000 = 0.00005$$

$$q = \sqrt{0.00005} = 0.007$$
$$= \text{about } 1/140 \text{ (frequency of recessive gene).}$$

The heterozygotes are represented by $2pq$ in the Hardy-Weinberg formula. Accordingly, the frequency of heterozygous carriers of albinism can be calculated as follows:

$$q = 0.007$$

$$p = 1 - 0.007 = 0.993$$

$$\therefore 2pq = 2\ (0.993 \times 0.007) = 0.014$$

$$= \text{about } 1/70 \text{ (frequency of heterozygote).}$$

Thus, although 1 person in 20,000 is an albino (recessive homozygote), about 1 person in 70 is a heterozygous carrier. There are 280 times as many carriers as affected individuals! It bears emphasizing that the rarity of a recessive disorder does not signify a comparable rarity of heterozygous carriers. In fact, when the frequency of the recessive gene is extremely low, nearly all the recessive genes are in the heterozygous state.

Selected Readings

Bodmer, W. F., and Cavalli-Sforza, L. L. 1976. *Genetics, evolution, and man.* San Francisco: W. H. Freeman and Co.

Dobzhansky, T. 1970. *Genetics of the evolutionary process.* New York: Columbia University Press.

Dunn, L. C., and Dobzhansky, T. 1952. *Heredity, race, and society.* New York: The New American Library.

Hardy, G. H. 1908. Mendelian proportions in a mixed population. *Science* 28: 49–50.

Li, C. C. 1955. *Population genetics.* Chicago: University of Chicago Press.

Li, C. C. 1961. *Human genetics*. New York: McGraw-Hill Book Co.

Mettler, L. E., and Gregg, T. G. 1969. *Population genetics and evolution*. Englewood Cliffs, N.J.: Prentice-Hall.

Stern, C. 1943. The Hardy-Weinberg law. *Science* 97: 137–38.

Stern, C. 1973. *Principles of human genetics*. San Francisco: W. H. Freeman and Co.

Wallace, B. 1968. *Topics in population genetics*. New York: W. W. Norton & Co.

5 Concept of Selection

We have already remarked that the multilegged anomaly, which appeared in 1958 in a local bullfrog population, has not been detected since that occasion. The supposition was made that the multilegged trait was governed by a recessive mutant gene. Some of us might presume that the mutant gene responsible for the abnormality has disappeared entirely from the population. But can a detrimental mutant gene be completely eradicated from natural populations of organisms, even in the face of the severest form of selection? Most persons are frankly puzzled when they are informed that the answer is *"No."* Yet, our knowledge of the properties of mutation and selection expressly permits a firm negative reply.

Selection Against Recessive Defects

In our consideration of the Hardy-Weinberg Law in the preceding chapter, we assumed that the mutant multilegged frogs (*aa*) were as reproductively fit as their normal kin (*AA* and *Aa*) and left equal numbers of living offspring each generation. Now, however, let us presume that all multilegged individuals fail to reach sexual maturity generation after generation. Will the incidence of the multilegged trait decline to a vanishing point?

We may start with the same distribution of individuals in the initial generation previously postulated in chapter 4, namely, 24 *AA*, 48 *Aa*, and 24 *aa*, with the sexes equally represented. Since the multilegged frogs (*aa*) are unable to participate in breeding, the parents of the next generation comprise only the 24 *AA* and 48 *Aa* individuals. The heterozygous types are twice as numerous as the homozygous dominants; accordingly, two-thirds of the total breeding members of the population are *Aa* and one-third are *AA*. We may once again employ a genetical checkerboard (table 5.1) to ascertain the different types of matings and their relative frequencies.

Table 5.1
Matings and Relative Frequencies

Female	Male (♂)	
(♀)	1/3 *AA*	2/3 *Aa*
1/3 *AA*	1/9 *AA* x *AA*	2/9 *AA* x *Aa*
2/3 *Aa*	2/9 *Aa* x *AA*	4/9 *Aa* x *Aa*

The frequencies of the different matings shown in table 5.1 may be expressed as whole numbers rather than fractions. Given a total of 36 matings, 4 (= 1/9 × 36) would be *AA* ♀ × *AA* ♂, 8 (= 2/9 × 36) would be *AA* ♀ × *Aa* ♂, 8 (= 2/9 × 36) would be *Aa* ♀ × *AA* ♂, and 16 (= 4/9 × 36) would be *Aa* ♀ × *Aa* ♂. These numbers are recorded in table 5.2.

Table 5.2
First Generation Offspring After Complete Selection

Type of Mating (Female x Male)	Number of Each Type of Mating*	Number of Offspring		
		AA	*Aa*	*aa*
AA x *AA*	4	16		
AA x *Aa*	8	16	16	
Aa x *AA*	8	16	16	
Aa x *Aa*	16	16	32	16
		64	64	16
		(44.44%)	(44.44%)	(11.11%)

* Based on a total of 36 matings.

Our next task is to determine the outcome of each type of cross. We shall assume that each mated pair contributes an equal number of progeny to the next generation (say, four offspring). As revealed in table 5.2, the offspring are distributed according to Mendelian ratios, and the actual numbers of offspring reflect the frequencies of the different kinds of matings. For example, a single *AA* ♀ × *Aa* ♂ mating yields 4 offspring in the Mendelian ratio of 2 *AA* : 2 *Aa*. There are, however, eight matings of this kind; the numbers of offspring are correspondingly increased to 16 *AA* and 16 *Aa*.

Even though all the multilegged frogs fail to reproduce, the detrimental recessive genes are still transmitted to the first generation. The emergence of multilegged frogs in the first generation stems from the matings of two heterozygous frogs. However, as seen from table 5.2, the frequency of the multilegged trait (*aa*) decreases from 25 percent to 11.11 percent in a single generation.

The effects of complete selection against the multilegged frogs in subsequent generations can be determined by the foregoing method of calculation, but the lengthy tabulations can be wearisome. At this point we may apply a formula that will establish in a few steps the frequency of the recessive gene after any number of generations of complete selection:

$$q_n = \frac{q_o}{1 + nq_o}.$$

In the above expression, q_o represents the initial or original frequency of the recessive gene, and q_n is the frequency after n generations. Thus, with the initial value of $q_o = 0.5$, the frequency of the recessive gene after two generations ($n = 2$) will be:

$$q_2 = \frac{q_o}{1 + 2q_o} = \frac{0.5}{1 + 2(0.5)} = \frac{0.5}{2.0} = 0.25.$$

If the frequency of the recessive gene itself (a) is q, then the frequency of the recessive individual (aa) is q^2. Accordingly, the frequency of the recessive homozygote is $(0.25)^2$, or 0.0625 (6.25 percent). In the second generation, therefore, the incidence of the multilegged trait drops to 6.25 percent.

If we perform comparable calculations through several generations, we emerge with a comprehensive picture that is tabulated in table 5.3 and portrayed in figure 5.1. In the third generation, the frequency of the recessive homozygote declines to 4.0 percent. Progress in the elimination of the multi-

Table 5.3
Effects of Complete Selection Against a Recessive Trait

Generations	Gene Frequency	Recessive Homozygotes %	Heterozygotes %	Dominant Homozygotes %
0	0.500	25.00	50.00	25.00
1	0.333	11.11	44.44	44.44
2	0.250	6.25	37.50	56.25
3	0.200	4.00	32.00	64.00
4	0.167	2.78	27.78	69.44
8	0.100	1.00	18.00	81.00
10	0.083	0.69	15.28	84.03
20	0.045	0.20	8.68	91.12
30	0.031	0.10	6.05	93.85
40	0.024	0.06	4.64	95.30
50	0.020	0.04	3.77	96.19
100	0.010	0.01	1.94	98.05

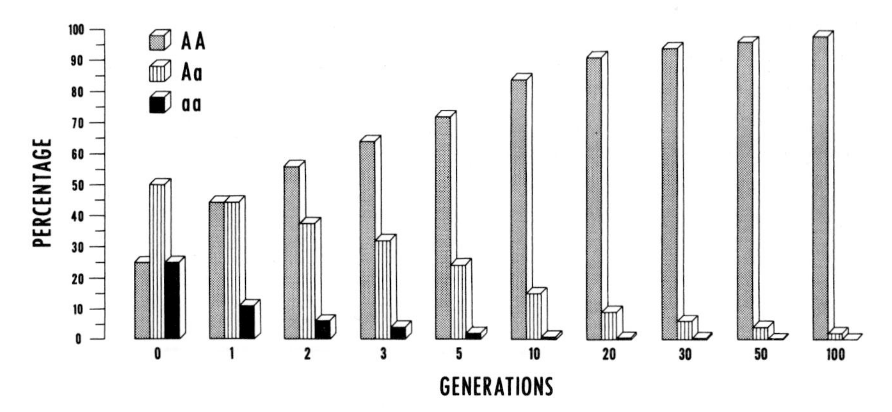

Figure 5.1 Effects of complete selection against recessive homozygotes (*aa*) occurring initially ("O" generation) at a frequency of 25 percent. The effectiveness of selection in reducing the incidence of the recessive trait decreases with successive generations. The frequency of recessive homozygotes drops markedly from 25 percent to 6.25 percent in two generations. However, 8 generations are required to reduce the incidence of the recessive trait to 1.0 percent, 30 generations are needed to achieve a reduction to 0.1 percent, and approximately 100 generations to depress the frequency to 0.01 percent.

legged trait is initially rapid, but becomes slower as selection is continued over many successive generations. About 20 generations are required to depress the incidence of the multilegged trait to 2 in 1,000 individuals (0.20 percent). Ten additional generations are necessary to effect a reduction to 1 in 1,000 individuals (0.10 percent). Thus, as a recessive trait becomes rarer, selection against it becomes less effective. The reason is quite simple: Only very few recessive homozygotes are exposed to the action of selection. The now rare recessive gene (*a*) is carried mainly by heterozygous individuals (*Aa*), where it is sheltered from selection by its normal dominant partner (*A*).

Significance of the Heterozygote

When the frequency of an abnormal recessive gene becomes very low, most affected offspring (*aa*) will come from matings of two heterozygous carriers (*Aa*). For example, in the human population, the vast majority of newly arising albino individuals (*aa*) in a given generation (more than 99 percent of them) will come from normally pigmented heterozygous parents. Considerations of this kind led us to postulate in chapter 1 that the multilegged frogs that appeared suddenly in the natural population were derived from normal-legged heterozygous parents (refer to fig. 1.4).

Detrimental recessive genes in a population are unquestionably harbored mostly in the heterozygous state. As shown in table 5.4, the frequency of heterozygous carriers is many times greater than the frequency of homozy-

Table 5.4
Frequencies of Recessive Homozygotes and Heterozygous Carriers

Frequency of Homozygotes (*aa*)	Frequency of Heterozygous Carriers (*Aa*)	Ratio of Carriers to Homozygotes
1 in 500 (Sickle-Cell Anemia)[a]	1 in 10	50:1
1 in 1,000 (Cystic Fibrosis)	1 in 16	60:1
1 in 6,000 (Tay-Sachs Disease)[b]	1 in 40	150:1
1 in 20,000 (Albinism)	1 in 70	285:1
1 in 25,000 (Phenylketonuria)	1 in 80	310:1
1 in 50,000 (Acatalasia)[c]	1 in 110	460:1
1 in 1,000,000 (Alkaptonuria)	1 in 500	2,000:1

[a] Based on incidence among American blacks.

[b] In the United States, the disease occurs once in 6,000 Jewish births and once in 500,000 non-Jewish births.

[c] Based on prevalence rate among the Japanese.

gous individuals afflicted with the trait. Thus, an extremely rare disorder, like alkaptonuria (blackening of urine), occurs in 1 in 1 million persons. This detrimental gene, however, is carried in the hidden state by 1 out of 500 persons. There are 2,000 as many genetic carriers of alkaptonuria as there are individuals afflicted with this defect. For another recessive trait, cystic fibrosis, 1 out of 1,000 individuals is affected with this homozygous trait. One of 16 persons is a carrier of cystic fibrosis. In modern genetic counseling programs, an important consideration has been the development of simple, inexpensive means of detecting heterozygous carriers of inherited disorders.

Interplay of Mutation and Selection

Theoretically, if the process of complete selection against the recessive homozygote were to continue for several hundred more generations, then the abnormal recessive gene would be completely eliminated, and the population would consist uniformly of normal homozygotes (AA). But, *in reality,* the steadily diminishing supply of abnormal recessive genes is continually being replenished by recurrent mutations from normal (A) to abnormal (a). Mutations from A to a, which inevitably occur from time to time, were not taken into account in our foregoing determinations. Mutations, of course, cannot be ignored.

All genes undergo mutation at some definable rate. If a certain proportion of A genes are converted into a alleles in each generation, then the population will at all times carry a certain amount of the recessive mutant gene (a) despite selection against it. Without any sophisticated calculations, it can be shown that a point will be reached at which the number of the abnormal recessive genes eliminated by selection just balances the number of the same recessive gene produced by mutation. An analogy shown in figure 5.2A will help in visualizing this circumstance. The water level in the beaker

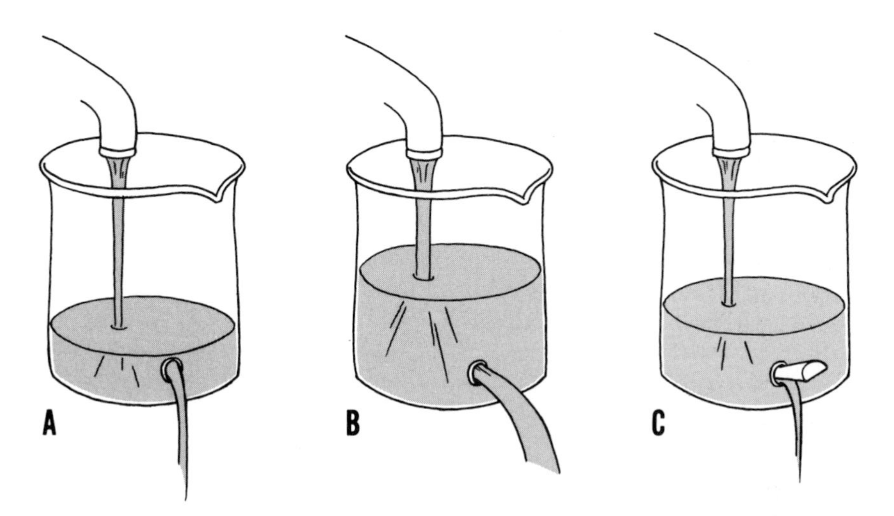

Figure 5.2 Interplay of detrimental mutant genes (water from faucet) and their elimination by selection (water escaping through hole) in a population (beaker) containing a pool of the harmful genes (water in beaker). *A*. State of genetic equilibrium (constant water level in beaker) when the rates at which water enters and leaves the beaker are equal. *B*. Effect of an increase in mutation rate (increased flow of faucet water) as might be expected from the continued widespread use of ionizing radiation. A new equilibrium (new constant water level) is established, but the frequency of the detrimental gene in the population is higher (higher water level in beaker). *C*. Effect of reducing selection pressure (decreased exit of water) as a consequence of improving the reproductive fitness of genetically defective individuals by modern medical practices. Mutation rate (inflow of water) is the same as in *A*. The inevitable result is a greater incidence of the harmful mutant gene in the population (higher level of water in the beaker).

remains constant when the rate at which water enters the opening of the beaker equals the rate at which it leaves the hole in the side of the beaker. In other words, a state of equilibrium is reached when the rate at which the recessive gene is replenished by mutation equals the rate at which it is lost by selection. It should be clear that it is not mutation alone that governs the incidence of deleterious recessives in a population. The generally low frequency of harmful recessive genes stems from the dual action of mutation and selection. The mutation process tends to increase the number of detrimental recessives; the selection mechanism is the counteracting agent. Mathematically, the frequency of the deleterious recessive gene at equilibrium can be calculated by the following formula, where q represents the frequency of the recessive gene and u represents the mutation rate of that gene:

$$q = \sqrt{u}.$$

What would be the consequences of an increase in the mutation rate? Man today lives in an environment in which high-energy radiation promotes a higher incidence of mutations. We may return to our analogy (fig. 5.2B). The increased rate of mutation may be envisioned as an increased input of water. The water level in the beaker will rise and water will escape more rapidly through the hole in the side of the beaker. Similarly, mutant genes will be found more frequently in a population, and they will be eliminated at a faster rate from the population. As before, a balance will be restored eventually between mutation and selection, but now the population has a larger store of deleterious genes and a larger number of defective individuals arising each generation.

The supply of defective genes in the human population has already increased through the greater medical control of recessive disorders. The outstanding advances in modern medicine have served to prolong the lives of genetically defective individuals who might otherwise not have survived to reproductive age. This may be compared to partially plugging the hole in the side of the breaker (fig. 5.2C). The water level in the beaker will obviously rise, as will the amount of deleterious genes in a population. Evidently, the price of our humanitarian principles is the enlargement of our pool of defective genes.

Partial Selection

We have treated above the severest form of selection against recessive individuals. Complete, or 100 percent, selection against a recessive homozygote is often termed *lethal* selection, and the mutant gene is designated as a *lethal* gene. A lethal gene does not necessarily result in the death of the individual, but does effectively prevent him from reproducing or leaving offspring. Not all mutant genes are lethal; in fact, the majority of them have less drastic effects on viability or fertility. A mildly handicapped recessive homozygote may reproduce but may be inferior in fertility to the normal individual. When the reproductive capacity of the recessive homozygote is only half as great as the normal type, he is said to be *semisterile,* and the mutant gene is classified as *semilethal.* A *subvital* recessive gene is one which, in double dose, impairs an individual to the extent that his reproductive fitness is less than 100 percent but more than 50 percent of normal proficiency.

The action of selection varies correspondingly with the degree of detrimental effect of the recessive gene. Figure 5.3 shows the results of different

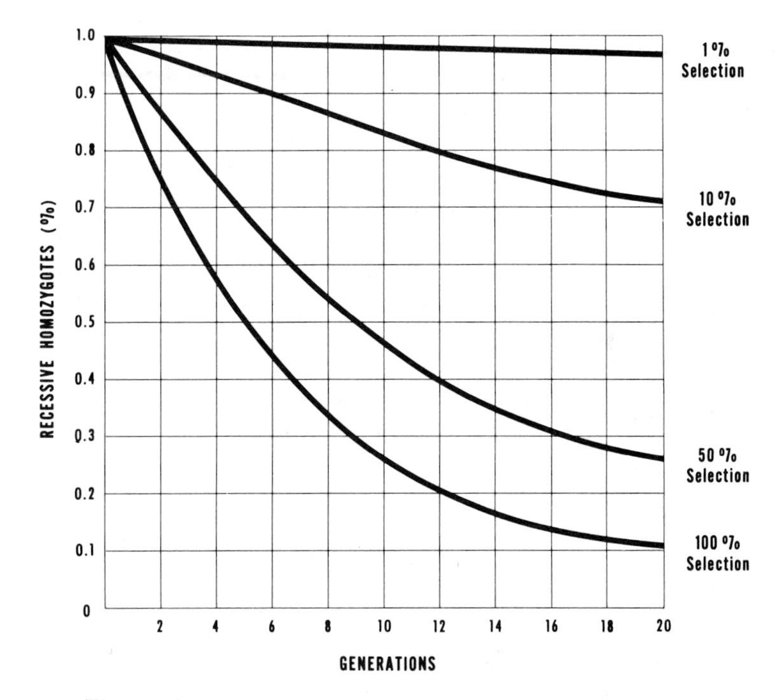

Figure 5.3 Different intensities of selection against recessive homozygotes occurring initially ("O" generation) at a frequency of 1.0 percent. The elimination of recessive individuals per generation proceeds at a slower pace as the strength of selection decreases.

intensities of selection in a population that initially contains 1.0 percent recessive homozygotes. With complete (lethal) selection, a reduction in the incidence of the recessive trait from 1.0 percent to 0.25 percent is accomplished in 10 generations. Twenty generations of complete selection reduces the incidence to 0.11 percent. When the recessive gene is semilethal (50 percent selection), 20 generations, or twice as many generations as under complete selection, are required to depress the frequency of the recessive homozygote to about 0.25 percent. Selection against a subvital gene (for example, 10 percent selection) results in a considerably slower rate of elimination of the recessive homozygotes. When the homozygote is at a very slight reproductive disadvantage (1.0 percent selection), only a small decline of 0.03 percent (from 1.0 percent to 0.97 percent) occurs after twenty generations. It is evident that mildly harmful recessive genes may remain in a population for a long time.

Selection Against Dominant Defects

If complete selection so acts against an abnormal trait caused by a dominant gene (*A*) that none of the *AA* or *Aa* individuals leave any progeny, then all the *A* genes are at once eliminated. In the absence of recurrent mutation, all subsequent generations will consist exclusively of homozygous recessives (*aa*).

However, we must contend again with ever-occurring mutations and the effects of partial selection. The geneticist Curt Stern, of the University of California, provides us with a simple, clear model of this situation. Imagine a population of 500,000 individuals, all of whom initially are homozygous recessive (*aa*). Thus, no detrimental dominant genes (*A*) are present and the population as a whole contains 1 million recessive genes (*a*). In the first generation, 10 dominant mutant genes arise, as a result of the recessive gene's mutating to the dominant state at a rate of 1 in 100,000 genes. We shall now assume that the dominant mutant gene is semilethal; in other words, only 5 of the newly arisen dominant genes are transmitted to the next, or second, generation. For ease of discussion, this is pictorially shown in figure 5.4 and represented also in table 5.5. It can be seen that the second

Table 5.5
Equilibrium Frequency of a Dominant Gene

Conditions:
1. Size of Population: 500,000 individuals
2. Mutation Rate ($a \longrightarrow A$): 1 in 100,000
3. Selection Coefficient (s) : 0.5 (semilethal)

Generation	Recessive Gene *a*	Dominant Gene *A*: Left over from Former Generations	Dominant Gene *A*: Newly Mutated	Dominant Gene *A*: Total
0	1,000,000	—	—	—
1	1,000,000*	—	10	10
2	1,000,000	5	10	15
3	1,000,000	5+2.5	10	17.5
4	1,000,000	5+2.5+1.25	10	18.75
5	1,000,000	5+2.5+1.25+0.625	10	19.375
∝	1,000,000	5+2.5+1.25+0.625+0.3125+...	10	20.00

* The total number of recessive genes should be reduced by the total number of dominant genes each generation, but this minor correction would be inconsequential.

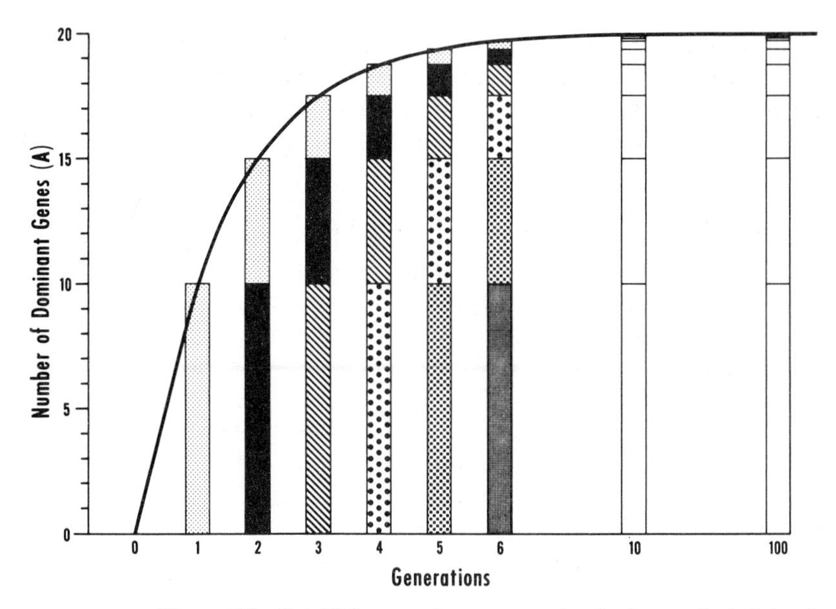

Figure 5.4 Establishment of a constant level of a semilethal dominant gene (A) in a population over the course of several generations. The fixed number of new dominant genes introduced each generation through mutations from a to A eventually exactly balances the number of dominant genes selectively eliminated each generation. In this particular case, an equilibrium is reached (after about 12 generations) when the total number of dominant genes is approximately 20.

generation would contain a total of 15 dominant genes—5 brought forth from the first generation and 10 new ones added by mutation. In the third generation, the 5 dominant genes carried over from the first generation would be reduced to 2.5, the 10 dominant alleles of the second generation would be depressed to 5, and 10 new abnormal alleles would arise anew by mutation. The total number of dominant genes would increase slightly each subsequent generation, until a point is reached (about 12 generations) where the rate of elimination of the abnormal dominant gene balances the rate of mutation. In other words, the inflow of new dominant alleles by mutation each generation is balanced by the outflow or elimination of the dominant genes each generation by selection.

The equilibrium frequency of the detrimental dominant gene in a population can be altered by changing the rate of loss of the gene in question. In man, retinoblastoma, or cancer of the eye in newborn babies, has until recently been a fatal condition caused by a dominant mutant gene. With modern medical treatment, approximately 70 percent of the afflicted individuals can be saved. The effect of increasing the reproductive fitness of the survivors is to raise the frequency of the abnormal dominant gene in the human population. The accumulation of detrimental genes in the human gene pool is a matter of growing awareness and concern.

Concealed Variability in Natural Populations

From what we have already learned, we should expect to find in natural populations a large number of deleterious recessive genes concealed in the heterozygous state. It may seem that this expectation is based more on theoretical deduction than on actual demonstration. This is not entirely the case. Penetrating studies by a number of investigators of several species of the fruit fly *Drosophila* have unmistakably indicated an enormous store of recessive mutant genes harbored by individuals in nature. We may take as an illustrative example the kinds and incidence of recessive genes detected in *Drosophila pseudoobscura* from California populations. The following data are derived from the studies of the noted geneticist Theodosius Dobzhansky, then at Columbia University but later at the University of California in Davis.

Flies were collected from nature, and a series of elaborate crosses were performed in the laboratory to yield offspring in which one pair of chromosomes carried an identical set of genes. The formerly hidden recessive genes in a given pair of chromosomes were thus all exposed in the homozygous state. All kinds of recessive genes were uncovered in different chromosomes, as exemplified by those unmasked in one particular chromosome, known simply as "the second." About 33 percent of the second chromosomes harbored one or more recessive genes that proved to be lethal or semilethal to flies carrying the second chromosome in duplicate. An astonishing number of second chromosomes—93 percent of them—contained genes that produced subvital or mildly incapacitating effects when present in the homozygous condition. Other unmasked recessive genes resulted in sterility of the flies or severely retarded the developmental rates of the flies. All these flies were normal in appearance when originally taken from nature. It is apparent that very few, if any, outwardly normal flies in natural populations are free of hidden detrimental recessive genes.

Genetic Load in Human Populations

The study of the concealed variability, or genetic load, in man cannot be approached, for obvious reasons, by the experimental breeding techniques used with fruit flies. Estimates of the genetic load in the human population have been based principally on the incidence of defective offspring from marriages of close relatives (consanguineous marriages). It can be safely stated that every human individual contains at least one newly mutated gene. It can also be accepted that any crop of gametes contains, in addition to one or more mutations of recent origin, at least 10 mutant genes that arose in the individuals of preceding generations and which have accumulated in the population. The average person is said to harbor four concealed lethal genes, each of which, if homozygous, is capable of causing death between birth and maturity. The most conservative estimates place the incidence of deformities to detrimental mutant genes in the vicinity of 2 per 1,000 births. *It is evident that man is not uniquely exempt from his share of defective genes.*

Muller's AID Plan

Modern medicine, by finding ways to keep alive individuals who carry deleterious genes, encourages the survival of those who more than likely will pass on their defects to future generations. The conventional ethics of medicine is shaken when, with increasing knowledge, it becomes clear that saving the life of a child with a hereditary disorder also ensures the retention and increase of detrimental genes that natural selection ordinarily keeps at very low frequencies. Can we continue indefinitely to load our population with hereditary disabilities?

Some geneticists have looked upon this situation with grave concern. In particular, the late Nobel laureate Hermann J. Muller was most distressed about the continual pollution of the human gene pool. Muller had been predicting genetic disaster since 1935, and throughout his career he was a persuasive and articulate prophet of doom. Muller presented the most vivid portrayal of the impending genetic disintegration of the human species. In the not too distant future, according to Muller, the task of taking care of genetically defective individuals will consume all the energy that society can mobilize. Most everyone will be an invalid. Muller's gloomy forecast was that the human species would end up with two types of individuals: one kind would be so genetically incapacitated as to be wheel-chair patients, and the other kind would be somewhat less disabled but would spend all their time taking care of the first kind. It is unreasonable, Muller contended, to expect medicine to keep up with the problem, especially because medical men themselves in that near or distant future will be subject to the same genetic decomposition. Eventually even the most sophisticated techniques available could no longer suffice to save men from their genetic deterioration. According to Muller, then, mankind is doomed unless positive steps are taken to regulate his genetic endowment.

Muller was convinced that man would be unable to reduce the load of unfavorable genes. He suggested as a countermeasure, therefore, a program designed to increase significantly the number of favorable genes in the human population. Muller proposed a plan called *AID,* or Artificial Insemination from Donors. He recommended the establishment of sperm banks, which would make available the sperm of highly qualified donors whose family histories showed the least possible likelihood of defects or abnormalities. To make sure of the eminence of the donors, the sperm would be frozen and made available only after 20 years or more, when the donor would no longer be alive, and posterity could judge dispassionately of his value. When a woman decided to have children, she need only choose sperm from the donor whose qualities she most admired. "How many women,"

Muller cried when he enunciated his plan, "would be eager and proud to bear and rear a child of Aldous Huxley or Charles Darwin!"

A number of serious questions must be answered before employing artificial insemination on the grand scale that Muller proposed. In the present state of knowledge even the best of geneticists would disagree on who should be the donors. There is no guarantee that even the most distinguished donor would be free of hidden detrimental genes. As one scientist stated, "The trouble with Muller's sperm bank is that he's always having to take people out of it." One of his rejects, for example, is Abraham Lincoln, who is now suspected of having suffered from a genetic disorder called Marfan's syndrome. Patients with Marfan's syndrome have poor musculature, long, thin extremities, and heart defects.

More importantly, people would be asked to surrender their pride in generating their own biological children and substitute for this pride the greater satisfaction to be gained from the knowledge that the child has the best possible set of genes. Muller confidently expected the gradual emergence of a new superior form of self-esteem and gratification in contributing to the genetic improvement of the human species. It seems more likely, however, that people would be rather uneasy at giving up their own claim to having their own biological descendants.

Other geneticists do not share Muller's pessimism that the human species is deteriorating genetically. Calculations indicate that the increase in the frequency of detrimental recessive genes due to the relaxation of selection by medical therapy would be *extremely slow*. For example, given the conditions presented in table 5.6, it would take well over 100 generations (or 3,000 years) for a recessive trait to double in frequency from 1 in 100,000

Table 5.6
Selection Relaxation for a Formerly Lethal Recessive Whose Fitness Is Restored to 100 Percent by Medical Therapy

Conditions: 1. Initial Frequency of Recessive Defect at Birth: 1/100,000 (1×10^{-5})
2. Mutation Rate ($A \longrightarrow a$): 1/100,000 (1×10^{-5})

Generation	Frequency of Recessive Defect at Birth
0	1×10^{-5}
1	1.01×10^{-5}
3	1.02×10^{-5}
10	1.06×10^{-5}
30	1.20×10^{-5}
100	1.73×10^{-5}

to 2 in 100,000. If the recessive defect occurred initially with a frequency of 1 in 10,000, then slightly more than 30 generations would be required to double the incidence. In essence, although relaxation of selection tends to engender an increase in the incidence of genetic defects in future generations, the increase occurs very slowly.

Selected Readings

Bajema, C. J., (ed.). 1971. *Natural selection in human populations.* New York: John C. Wiley & Sons.

Cavalli-Sforza, L. L., and Bodmer, W. F. 1971. *The genetics of human populations.* San Francisco: W. H. Freeman and Co.

Crow, J. F. 1961. Population genetics. *American Journal of Human Genetics* 13: 137–50

Dobzhansky, T. 1950. The genetic basis of evolution. *Scientific American,* January, pp. 32–41.

Dobzhansky, T. 1961. Man and natural selection. *American Scientist* 49:285–99.

Hardin, G. 1959. *Nature and man's fate.* New York: The New American Library.

Morris, L. N. 1971. *Human populations, genetic variation, and evolution.* New York: Chandler Publishing Co.

Neel, J. V., and Schull, W. J. 1968. On some trends in understanding the genetics of man. *Perspectives in Biology and Medicine* 11:565–602.

Reed, S. C. 1964. *Parenthood and heredity.* New York: John C. Wiley & Sons.

Wallace, B. 1970. *Genetic load.* Englewood Cliffs, N.J.: Prentice-Hall.

6

Selection in Action

The process of evolution is slow and continuous. We have seen that many generations of persistent selection are required to reduce drastically the frequency of an unfavorable mutant gene in a population. Likewise, it generally takes an inordinately long period of time for a new favorable mutant gene to replace its allele throughout a large population. Yet, we have encountered situations in nature in which a favorable mutation has spread through a population in a comparatively short span of years. We shall look at some outstanding examples in which man has actually observed evolution in progress.

Industrial Melanism

One of the most spectacular evolutionary changes witnessed by man has been the emergence and predominance in modern times of dark, or *melanic,* varieties of moths in the industrial areas of England and continental Europe. Slightly more than a century ago dark-colored moths were exceptional. The typical moth in the early 1800s had a light color pattern, which blended with the light coloration of tree trunks on which the moths alighted. But then the industrial revolution intervened to alter materially the character of the countryside. As soot and other industrial wastes poured over rural areas, the vegetation became increasingly coated and darkened by black smoke particles. In areas heavily contaminated with soot, the formerly abundant light-colored moths have been supplanted by the darker varieties. This dramatic change in the coloration of moths has been termed "industrial melanism." At least 70 species of moths in England have been so affected by man's disturbance of the environment.

During the past two decades, several scientists, particularly E. B. Ford and H. B. D. Kettlewell at the University of Oxford, have analyzed the phenomenon of industrial melanism. Kettlewell photographed the light and dark forms of the peppered moth, *Biston betularia,* against two different backgrounds (fig. 6.1). The light variety is concealed and the dark form is clearly visible when the moths rest on a light lichen-coated trunk of an oak tree in an unpolluted rural district. Against a sooty black oak trunk, the light form is conspicuous and the dark form is well camouflaged. Records of the dark form of the peppered moth date back to 1848, when its occurrence was reported at Manchester in England. At that time, the dark form comprised less than 1 percent of the population. By 1898, only 50 years later, the dark form had come to dominate the Manchester locale, having attained a remarkably high frequency of occurrence estimated at 95 percent. In fact, the incidence of the melanic type has reached 90 percent or more in most British industrial areas.

The rapid spread of the dark variety of moth is quite evident. The dark variants are protectively colored in the smoke-polluted industrial regions. They more easily escape detection by predators—insect-eating birds. Actual films taken by Kettlewell and Niko Tinbergen reveal that birds prey on the moths in a selective manner. That is, predatory birds more often capture the conspicuous light-colored moths in polluted woodlands. In a single day, the numbers of light forms in an industrial area may be reduced by as much as one-half by bird predation.

Figure 6.1 Dark and light forms of the peppered moth (*Biston betularia*) clinging to a soot-blackened oak tree in Birmingham, England (left) and to a light, lichen-coated oak tree in an unpolluted region (*right*). (Courtesy of Dr. H. B. D. Kettlewell.)

Experimental breeding tests have demonstrated that the two varieties differ principally by a single gene, with the dark variant being dominant to the light one. The dominant mutant gene was initially disadvantageous and probably was maintained at an extremely low frequency in nature by a balance between mutation and selection (see chapter 5). Then, as an indirect consequence of industrialization, the mutant gene became favored by natural selection and spread rapidly in populations in a comparatively short period of time. In unpolluted or nonindustrial areas in western England and northern Scotland, the dominant mutant gene does not confer an advantage on its bearers, and the light recessive moth remains the prevalent type.

One of the many impressive features of Kettlewell's studies lies in the unequivocal identification of the selecting agent. Selection, we may recall, has been defined as differential reproduction. The act of selection in itself does not reveal the factors or agencies that enable one genotype to leave more offspring than another. We may demonstrate the existence of selection, yet not know the cause of that selection. We might have reasonably suspected that predatory birds were directly responsible for the differential success of the melanic forms in survival and reproduction, but Kettlewell's laboriously accumulated data provided that all-important, often elusive, ingredient: *proof*.

If the environment of the peppered moth were to become altered again, then natural selection would be expected to favor the light variety again. In the 1950s, the British Parliament passed the Clean Air Act, which decreed, among other things, that factories must switch from soft high-sulfur (sooty) coal to less smoky fuels. The enforcement of this enlightened smoke-abatement law has led to a marked reduction in the amount of soot in the atmosphere. In the 1970s, the University of Manchester biologist L. M. Cook and his colleagues reported a small, but significant, increase in the frequency of the light-colored peppered moth in the Manchester area. This is further substantiation of the action and efficacy of natural selection.

Australian Rabbits

The European wild rabbit, *Oryctolagus cuniculus,* has gained infamous notoriety in Australia as the most serious economic pest ever introduced on this isolated island continent. In 1859, a small colony of 24 wild rabbits was brought from Europe to an estate in Victoria in the southeastern corner of Australia. From such modest beginnings, the rabbits multiplied enormously and by 1928 had spread over the greater part of the Australian continent. Estimates placed the number of adult rabbits at over 500 million in an area of about 1 million square miles. In overrunning the open grassy plains, the rabbits caused extensive deterioration to sheep-grazing pastures and to wheat fields.

For many years, the Australian government spent large sums of money on various measures to control the population explosion of these prolific rabbits. Trapping, rabbit-proof fencing, poisoning of water holes, and fumigation all proved to be largely ineffectual. Then, beginning in 1950, outstanding success in reducing the rabbit population was achieved by inoculating rabbits with a virus that causes the fatal disease *myxomatosis*. The deadly myxoma virus was implanted into the tissues of rabbits in the southern area of Australia (fig. 6.2). In a remarkably short period of time, the

Figure 6.2 Typical lesion that develops in a rabbit infected with myxoma virus. (Courtesy of Dr. Frank J. Fenner.)

virus had made its way, aided by insect carriers (mosquitoes), into most of the rabbit-infested areas of the continent. By 1953, more than 95 percent of the rabbit population in Australia had been annihilated.

However, after their drastic decline in the early 1950s, the rabbit populations began to build up again. Evolutionary changes have occurred in both the pathogen (virus) and the host (rabbit). Mutations conferring resistance to the myxoma virus have selectively accumulated in the rabbit populations. At the same time, the viruses themselves have undergone genetic changes; less virulent strains of the virus have evolved.

Selection would favor the emergence of less virulent strains of a pathogen leading to less severe expressions of the disease. The advantage to the pathogen is comprehensible, since a pathogen which would cause the instantaneous death of the host also would cause its own prompt demise from lack of an immediate host. Experiments by the Australian scientist Frank J. Fenner in the 1950s showed that new strains of the myxoma virus had arisen. Highly susceptible, standard laboratory rabbits were infected with different preparations that had been isolated in successive epidemics in Australia. The early, or original, virus strains caused greater mortality in the laboratory rabbits than strains taken from the later epidemics. With each successive epidemic, a less virulent strain had evolved. In nature, the original, highly virulent strain of virus had killed the rabbits within a few days, which limited the virus' chances of being transmitted to a new host through the mosquito vector. However, when a virus of lower virulence appeared, the duration of the disease in the rabbit was extended, with the consequence that the less virulent mutant strain improved its chances of being transmitted to unaffected rabbits through the bite of a mosquito. Accordingly, attenuated, or weakened, strains of the virus progressively displaced in nature the original highly virulent strain.

Experiments by Fenner and his collaborators also demonstrated that the rabbits had become genetically resistant to the virus. Uniform doses of a standard virus preparation were injected into laboratory offspring of rabbits that had been trapped in successive epidemics. Significantly greater numbers of experimental offspring of parents from the first drastic epidemic succumbed to the inoculated virus compared with offspring of parents from the later, relatively mild epidemics. Thus, successive epidemics fostered the accumulation of genes protecting the rabbits against death from the myxoma virus. Stated another way, genetically susceptible rabbits had died, leaving the more genetically resistant rabbits to contribute more offspring each generation. In essence, natural selection has favored a mutual accommodation between the virus and the Australian rabbit.

Selection for Resistance

Penicillin, sulfonamides (sulfa drugs), streptomycin, and other modern antibiotic agents made front-page headlines when first introduced. These wonder drugs were execeptionally effective against certain disease-producing bacteria, and contributed immeasurably to the saving of human lives in World War II. However, the effectiveness of these drugs has been reduced by the emergence of resistant strains of bacteria. Medical authorities regard the rise of resistant bacteria as the most serious development in the field of infectious diseases over the past decade. Bacteria now pass on their resistance to antibiotics faster than people spread the infectious bacteria.

Mutations have occurred in bacterial populations that enable the mutant bacterial cells to survive in the presence of the drug. Here again we notice that mutations furnish the source of evolutionary changes, and that the fate of the mutant gene is governed by selection. In a normal environment, mutations that confer resistance to a drug are rare or undetected. In an environment changed by the addition of a drug, the drug-resistant mutants are favored and supplant the previously normal bacterial strains.

It might be thought that the mutations conferring resistance are actually caused or induced by the drug. But this is not the case. Drug-resistant mutations arise in bacterial cells irrespective of the presence or absence of the drug. An experiment devised by the Stanford University geneticist Joshua Lederberg provides evidence that the drug acts as a selecting agent, permitting preexisting mutations to express themselves. As seen in figure 6.3, colonies of bacteria were grown on a streptomycin-free agar medium in a petri plate. When the agar surface of this plate was pressed gently on a piece of sterile velvet, some cells from each bacterial colony clung to the fine fibers of the velvet. The imprinted velvet could now be used to transfer the bacterial colonies onto a second agar plate. In fact, more than one replica of the original bacterial growth can be made by pressing several agar plates on the same area of velvet. This ingenious technique has been appropriately called *replica plating.*

In preparing the replicas, Lederberg used agar plates containing streptomycin. Of course, on these agar plates, only bacterial colonies resistant to streptomycin grew. In the case depicted in figure 6.3, one colony was resistant. Significantly, this one resistant colony was found in the same exact position in all replica plates. If mutations arose in response to exposure to a drug, it is hardly to be expected that mutant bacterial colonies would arise in precisely the same site on each occasion. In other words, a haphazard or random distribution of resistant bacterial colonies, without restraint or attention to location in the agar plate, would be expected if the mutations did not already exist in the original bacterial colonies.

Now, we can return to the original plate, as Lederberg did, and test samples of the original bacterial colonies in a test tube for sensitivity or resistance to streptomycin (bottom part of fig. 6.3). It is noteworthy that the bacterial colonies on the original plate had not been previously in contact with the drug. When these original colonies were isolated and tested for resistance to streptomycin, only one colony proved to be resistant. This one colony occupied a position on the original plate identical with the site of the resistant colony on the replica plates. The experiment demonstrated conclusively that the mutation had not been induced by streptomycin but had already been present before exposure to the drug.

Other genetic changes perpetuated by selection have been detected in recent years. The insecticide with the awesome name of dichloro-diphenyl-trichloroethane, or DDT, was initially highly poisonous to malaria-carrying mosquitoes and typhus-carrying lice. However, in the 27 years since the introduction of this insect poison, numerous mutant DDT-resistant strains of mosquitoes and lice have appeared in widely different parts of the world. Similarly, hydrogen cyanide has for many years been an efficient fumigant in the destruction of insect pests, particularly the scale insect, in the citrus-producing areas of southern California. Mutant strains of scale insects that can tolerate the usual lethal doses of hydrogen cyanide have recently arisen in several of the fruit orchards of California. Man's struggle against diseases

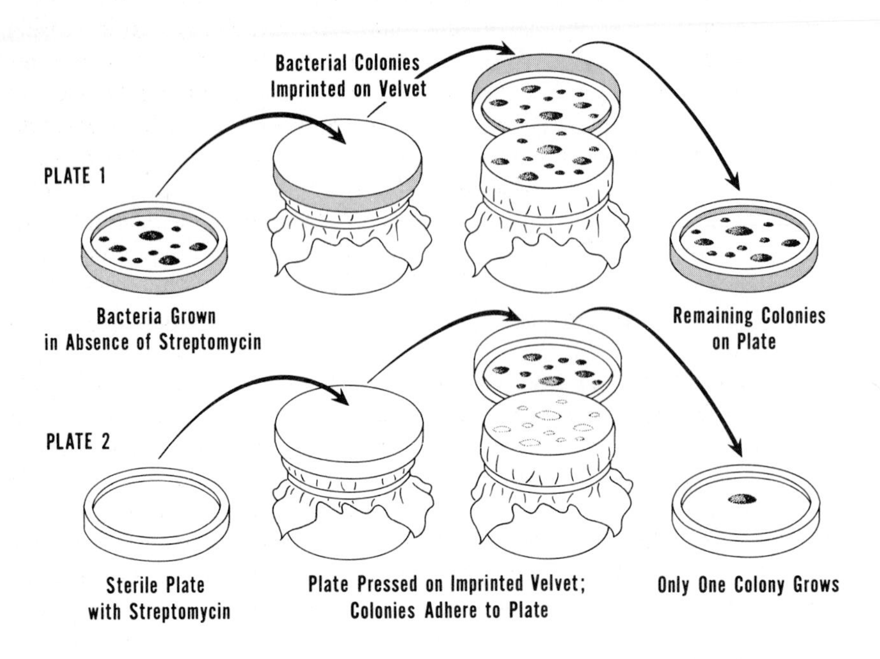

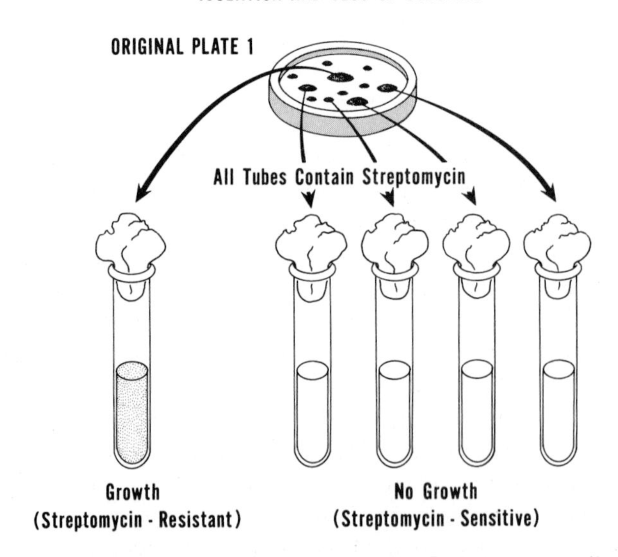

Figure 6.3 Experiment by Joshua Lederberg revealing that drug-resistant mutations in bacterial cells had not been induced by the drug but had already been present prior to exposure of the bacteria to the drug.

and insect pests is unending as long as the mutation process continually produces variants that may have survival value in an altered environment.

Parenthetically, it should be noted that DDT has fallen into increasing disfavor as a threat to wildlife. DDT is an extremely stable compound, retaining its chemical potency for many years rather than breaking down harmlessly within a few days after application. Long-lasting DDT residues have spread within two decades throughout the world by wind and water in much the same way as radioactive fallout. DDT has been found in areas of the globe where the insecticide has never been used. Penguins and crab-eating seals in the Antarctic, far removed from any actual spraying, have accumulated traces of the compound. It has been estimated that there are approximately one billion pounds of DDT currently circulating in the biosphere.

The DDT levels in the gonads and eggs of several species of birds have seriously impaired the reproductive potential of these organisms. Since DDT interferes with the mobilization of calcium in the oviduct, the eggs are laid with exceedingly thin shells, subject to hairline cracks, or with no shell at all. Birds of prey, such as the bald eagle and osprey, are now on the brink of extinction. Whereas ospreys formerly averaged 2.5 young a year, today they are barely able to produce 1 young a year. As William Mudoch of the University of California (Santa Barbara) stated, "no amount of technological ingenuity can reconstruct a species of osprey once it is extinct." Even with the curbed use of DDT in several countries (Sweden, Canada, and the United States), it would take several decades for the environment to purge itself of this extremely long-lasting pollutant.

Infectious Diseases in Man

The British geneticist J. B. S. Haldane has suggested that infectious diseases have been one of the most potent agents of natural selection in human populations. It is difficult to prove unequivocally that genetic changes have occurred in the resistance of man to his pathogens during widespread epidemics in the past. Nevertheless, it is a reasonable supposition that major epidemics eliminated a large percentage of genetically susceptible individuals, and permitted the selective survival and multiplication of individuals endowed with greater genetic resistance.

Up until 10,000 years ago, man subsisted by food-gathering and hunting alone. Widely dispersed groups of nomadic peoples lived at low population densities. Contact between groups was limited, and contagious diseases occurred infrequently. With the development of agriculture and the establishment of village settlements, human populations became dense and concentrated in relatively small regions. This set the stage for infectious community diseases, spread by contact, such as tuberculosis and bubonic plague, both caused by bacteria. Viral diseases such as measles, mumps, chickenpox, and smallpox posed major threats to the health of preindustrial man. With the advent of the industrial revolution, the densely populated urban areas witnessed repeated outbreaks of infectious diseases. In industrial times, many diseases—in particular, tuberculosis—flared up to ravage entire populations.

An infectious disease tends to reach epidemic proportions in populations that have had no previous exposure to the disease. When populations are first exposed to a contagious disease, the mortality is high and the infection is acute. On evolutionary principles, repeated exposure over many generations should lead to a genetic reconstruction of the population, with a raised frequency of genetically resistant individuals. Thus, in subsequent episodes of the disease, the expectations are lowered mortality and less severe manifestations of the disease.

Strong presumptive evidence that heritable resistance can arise through selection is furnished by the change, with time, in the incidence and character of tuberculosis among the Plains Indians of the Qu'Appelle Valley Reservation in Saskatchewan, Canada. In 1881, the Plains Indians, shortly after arriving at the reservation, became exposed to the tuberculosis germ for the first time. By 1886, almost everyone in the reservation had contracted the infection. The annual death rate from tuberculosis in 1886 reached the incredible figure of 900 per 10,000, or approximately 10 percent of the total population. The disease was not only widespread but also highly virulent, causing damage to bone, meninges, kidney, and other parts of the body. Such multiple disorders reflect the inability of the host to localize the infection. In 1921, however, the disease showed a pronounced tendency to localize in the lungs and to pursue the chronic course characteristic of pulmonary tuberculosis. Mortality dropped to 7 percent among Indian school children in 1921, and in 1950, mortality declined to 0.2 percent. Accordingly, in but a few generations, less than one percent of the Indian children died of tuberculosis. It should be stressed that this dramatic decline occurred in absence of specific chemotherapy and modern public health measures.

Other communities of the world have shown the same pattern of evolutionary change. Europeans typically exhibit a mild form of tuberculosis. René Dubos, a microbiologist at the Rockefeller Institute, carefully analyzed the available data, and concluded that Europeans today are the beneficiaries of an exacting selective process brought about by repeated outbreaks of tuberculosis a few generations ago. During the 18th and 19th centuries, tuberculosis spread with the growth of cities, and created havoc throughout the industrialized world. It became known as the "Great White Plague." Annual mortality rates during the industrial era were as high as 500 per 100,000.

Tuberculosis is a highly contagious disease. Minute droplets of sputum, discharged by cough or sneeze of an affected person, contain literally thousands of tubercle bacteria, which can survive in the air for several hours. Even thoroughly dried sputum, when it is inhaled, can be infective. Tuberculosis flared in crowded London in 1750, and then swept uncontrollably through the large cities of continental Europe. Immigrants from Europe brought the devastating disease to the United States in the early 1800s. The death toll was highest in cities with large immigrant populations—New York, Philadelphia, and Boston. As tuberculosis advanced across the nation, it became the chief cause of death in the United States in the 19th century. Tuberculosis then declined steadily in the 20th century in all industrialized countries. From a high value of 500 deaths per 100,000 in the 19th century, the mortality rate dropped in Europe and the United States to approximately 190 per 100,000 in 1900. By 1945, the rate fell to about 40 per 100,000.

The decline in mortality cannot be attributable, in any large measure, to general advances in living conditions, or to therapy. The decline was in full force before effective medical steps were taken, and indeed, even before the discovery of the tubercle bacteria. The causative agent was discovered only in 1882 by Robert Koch in Germany. The available information suggests strongly that the reduction in mortality has been primarily the result of an evolutionary attenuation of the virulence of the pathogen coupled with a heightened, genetically conditioned, resistance level of the host. In other words, the severity of infection has most likely been diminished by genetic selection and has been essentially independent of medical or public health intervention.

It is only since the 1940s that the benefits of chemotherapy have become detectable in the Western world. Today, cases of tuberculosis occur at an annual rate of 20 per 100,000, and most cases are permanently arrested by treatment with drugs such as isonicotinic acid hydrazide (INH) and para-amino salicylic acid (PAS). The effectiveness of modern drugs seems to indicate that populations of parts of the world in which tuberculosis still prevails can avoid the remorseless selective process in which death eliminates the susceptible genotypes. Yet, some 3 million persons each year succumb to tuberculosis in Latin America, Africa, and Asia. Although effective drugs to combat tuberculosis are available for worldwide distribution, most of the poorly developed nations have inadequate medical and health services, and large proportions of their populations have only remote contact with any medical care.

As the prevalence of tuberculosis has decreased with the use of chemotherapeutic agents, so has the threat to genetically susceptible individuals. Persons who earlier would have succumbed to tuberculosis are now surviving. In other words, the reproductive fitness of genetically susceptible individuals has increased. Reduction in selection pressure for disease resistance has probably been accompanied by an increase in the frequency of genetic factors for susceptibility to tuberculosis. Since other animal species continue to harbor the tuberculosis germ, human populations would be especially vulnerable to outbreaks of tuberculosis in the event of lapses in public health measures.

Selected Readings

Bishop, J. A., and Cook, L. M. 1975. Moths, melanism and clean air. *Scientific American,* January, pp. 90–99.

Dubos, R. 1968. *So human an animal.* New York: Charles Scribner's Sons.

Dubos, R., and Dubos, J. 1952. *The white plague: tuberculosis, man and society.* Boston: Little, Brown and Co.

Fenner, F. 1959. Myxomatosis in Australian wild rabbits—evolutionary changes in an infectious disease. *The Harvey Lectures 1957–58.* New York: Academic Press.

Ford, E. B. 1960. *Mendelism and evolution.* London: Methuen & Co.

Harrison, G., Weiner, J., Tanner, J., and Barnicot, N. 1964. *Human biology: an introduction to human evolution.* New York: Oxford University Press.

Huxley, J. 1953. *Evolution in action.* New York: New American Library of World Literature.

Kettlewell, H. B. D. 1959. Darwin's missing evidence. *Scientific American,* March, pp. 48–53.

Motulsky, A. G. 1960. Metabolic polymorphisms and the role of infectious diseases in human evolution. *Human Biology* 32:28–62.

Neel, J. V. 1958. The study of natural selection in primitive and civilized human populations. *Human Biology* 30:43–72.

Ryan, F. J. 1953. Evolution observed. *Scientific American,* October, pp. 78–83.

Tax, S. ed. 1960. *Evolution after Darwin* (3 vols.). Chicago: University of Chicago Press.

7 Balanced Polymorphism

The concepts presented in the preceding chapters have led us to believe that selection operates at all times to reduce the frequency of an abnormal gene to a low equilibrium level. This view is not entirely accurate. We are aware of genes with deleterious effects that occur at fairly high frequencies in natural populations. A striking example is the high incidence in certain human populations of a mutant gene that causes a curious and usually fatal form of blood cell destruction, known as *sickle-cell anemia.* It might be presumed that this detrimental gene is maintained at a high frequency by an exceptionally high mutation rate. There is, however, no evidence to indicate that the sickle-cell gene is unusually mutable. We now know that the maintenance of deleterious genes at unexpectedly high frequencies involves a unique, but not uncommon, selective mechanism, which results in a type of population structure known as *balanced polymorphism.*

Sickle-Cell Anemia

This disease was discovered by the American physician James B. Herrick, who in 1904 made an office examination of an anemic West Indian black student residing in Chicago. The patient's blood examined under the microscope showed the presence of numerous crescent-shaped erythrocytes. The peculiarly twisted appearance of the red blood cell is shown in figure 7.1. The black patient was kept under observation for six years, during which time he displayed many of the distressing symptoms we now recognize as typical of the disease. The bizarre-shaped red cells in the form of a sickle blade are fragile and clog small blood vessels. The obstruction of circulation leads in turn to the necrosis (death) of various tissues. The clogging of small blood vessels is responsible for the localized painful "crises" of sickle-cell anemia. The victim may suffer from pneumonia as a result of lung damage,

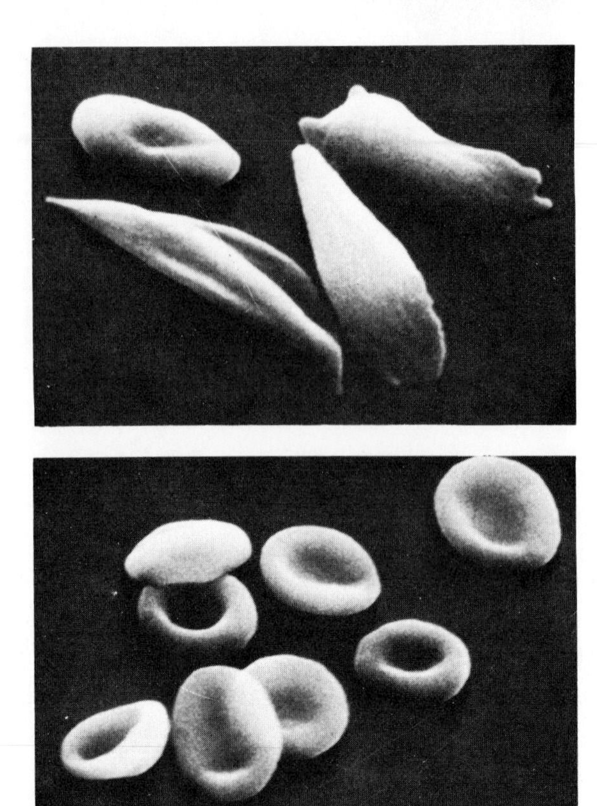

Figure 7.1 Peculiarly twisted shape of red blood cells of an individual suffering from sickle-cell anemia (*top*), contrasted with spherical appearance of normal red blood cells (*bottom*). (Courtesy of Wide World Photos.)

rheumatism from muscle and joint deterioration, heart damage, and kidney failure.

Affected persons rarely reach adult life; most die in the first decade and very few survive the third. Physicians can provide care for the patient, but there is no cure. In other words, there is little that can be done for patients in sickle-cell crisis beyond treatment of the symptoms. In 1970, Robert Nalbandian, a pathologist in Grand Rapids, Michigan, reported success in desickling afflicted blood cells by injecting patients with large amounts of *urea*. The effect of urea is transient, and the efficacy of urea therapy awaits clarification. Anthony Cerami and James Manning at the Rockefeller University in New York have proposed that the agent that relieves sickle-cell crisis in the capillaries is not the urea itself, but an impurity, *cyanate,* present in small amounts in most urea solutions. The antisickling effect of cyanate appears to be more enduring, and only small doses of cyanate are required to alleviate the sickle-cell crisis.

Sickle-cell anemia occurs predominantly in blacks. It is about six times more common than the next most common long-term illness of black children (diabetes). The incidence at birth of the disabling sickle-cell anemia in the United States is estimated at 2 per 1,000 infants. Not until after World War II was the hereditary basis of sickle-cell anemia uncovered. The irregularity was shown in the late 1940s by James V. Neel of the University of Michigan to be inherited as a simple Mendelian character. The sickle-cell anemic patient inherits two defective genes, one from each parent (fig. 7.2). Individuals with one normal and one defective (sickling) gene are generally healthy but are carriers—they have *sickle-cell trait.* If two heterozygous carriers marry, the chances are one in four that a child will have sickle-cell anemia, and one in two that a child will be a carrier (fig. 7.2).

Since the homozygous state of the detrimental gene is required for the overt expression of the disease, sickle-cell anemia may be considered to be recessively determined. However, although heterozygous individuals are, on the whole, normal, even the red cells of the heterozygotes can undergo sickling under certain circumstances, producing clinical manifestations. As an instance, heterozygous carriers have been known to experience acute abdominal pains at high altitudes in unpressurized airplanes. The lowering of the oxygen tension, with an increased amount of oxygen released from chemical combination in the erythrocytes, is sufficient to induce sickling. The abdominal pains can be traced to the packing of sickled erythrocytes in the small capillaries of the spleen.

According to some authors, since the detrimental gene can express itself in the heterozygous state (by producing a positive sickling test under certain circumstances), the gene should be considered dominant. This reveals that "dominance" and "recessiveness" are somewhat arbitrary concepts that depend on one's point of view. Indeed, from a molecular standpoint, the relation between the normal and defective gene in this instance may best be described as *codominant,* since, as we shall see below, the heterozygote produces both normal and abnormal hemoglobin. There is no blending of inheritance at the molecular level.

Sickle-Cell Hemoglobin

In 1949, the distinguished chemist and Nobel laureate Linus Pauling and his co-workers made the important discovery that the defective sickling gene alters the configuration of the hemoglobin molecule. Pauling used the then relatively new technique of electrophoresis, which characterizes proteins according to the manner in which they move in an electric field. The hemoglobin molecule travels toward the positive pole. Hemoglobin from a sickled cell differs in speed of migration from normal hemoglobin; it moves more slowly than the normal molecule (fig. 7.3). The defective gene thus functions differently from the normal gene, and, in fact, acts independently of its normal partner allele. As a result, the heterozygote does not produce an intermediate product, but instead produces both kinds of hemoglobin in

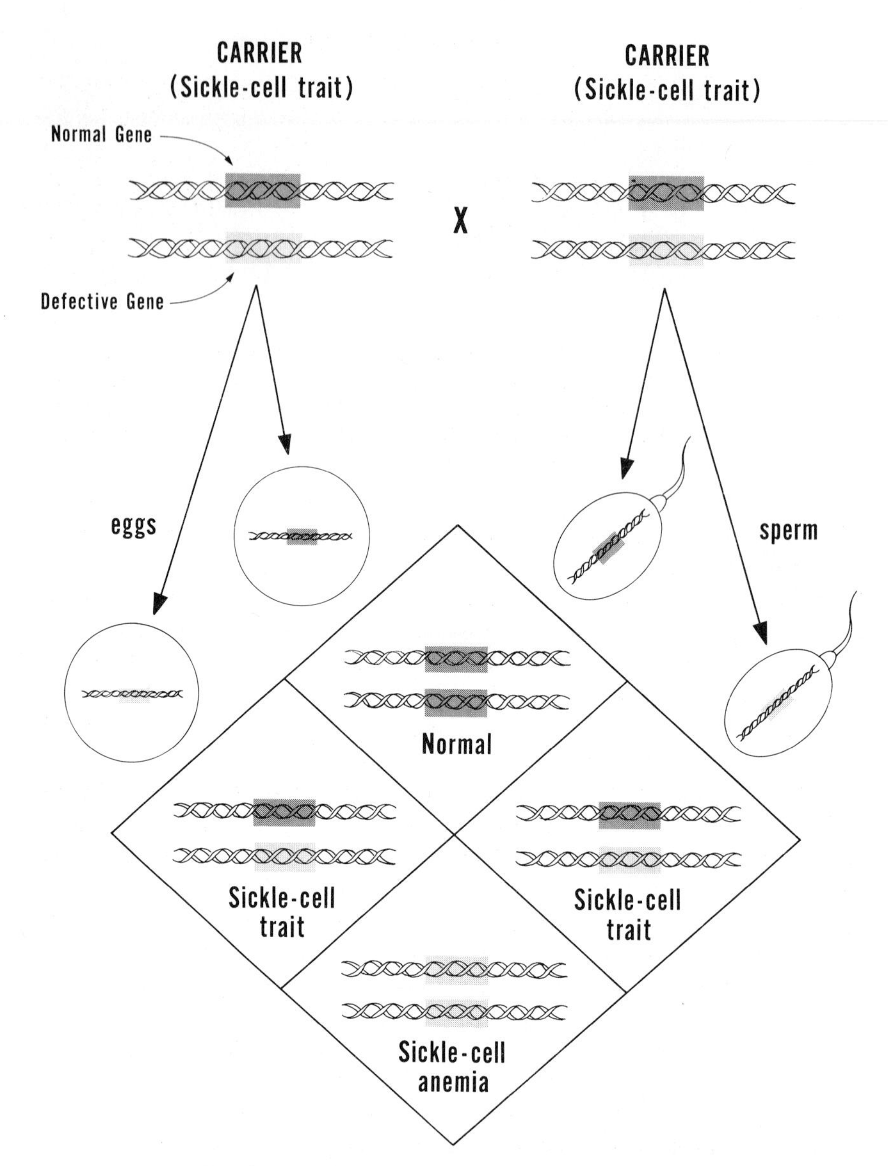

Figure 7.2 Types of children that can result from a marriage of two heterozygous carriers of sickle-cell anemia. Individuals homozygous for the variant gene suffer from sickle-cell anemia; the benign heterozygous state is referred to as the sickle-cell trait.

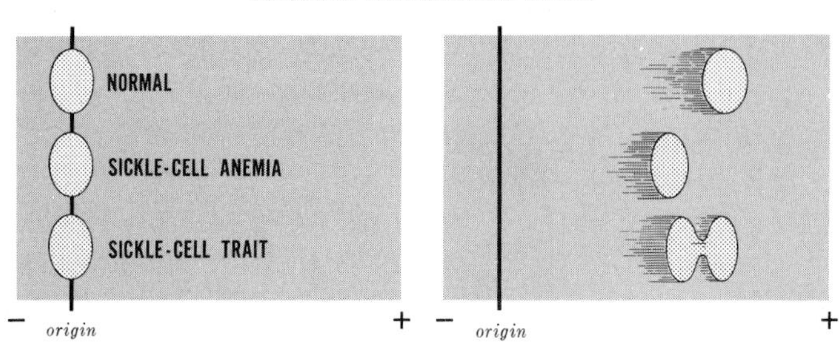

Figure 7.3 Electrophoretic patterns of hemoglobins. The hemoglobin of the heterozygous person with sickle-cell trait is not intermediate in character, but is composed instead of approximately equal proportions of the normal hemoglobin and the sickle-cell anemia variety. (Based on studies by Linus Pauling.)

nearly equal quantities—the normal type (designated *hemoglobin A,* or *Hb A*) and the sickle-cell anemic variety (*Hb S*). The dual electrophoretic pattern of hemoglobin from a heterozygous individual can actually be duplicated experimentally by mechanically mixing the hemoglobins taken from blood cells of a normal person and a sickle-cell anemic patient. The mixed solution separates in an electric field into the same two hemoglobin components as those characteristic of a heterozygous person with sickle-cell trait.

The sickle-cell mutant gene modifies the molecular structure of hemoglobin, and sickle-cell anemia has been appropriately described as a *molecular disease.* It remained for the chemist Vernon Ingram, then at Cambridge University in England, to discover how the hemoglobin molecule is altered by the aberrant gene. In 1956, Ingram succeeded in breaking down hemoglobin, a large protein molecule, into several peptide fragments containing short sequences of identifiable amino acids. Normal hemoglobin and sickle-cell hemoglobin yielded the same array of peptide fragments, with a single exception. In one of the peptide fragments of sickle-cell hemoglobin, the amino acid *glutamic acid* had been replaced at one point in the chain by *valine* (fig. 7.4). The sole difference in chemical composition between normal and sickle-cell hemoglobin is the incredibly minute substitution of a single amino acid unit among several hundred. The fatal effect of sickle-cell anemia is thus traceable to an exceedingly slight alteration in the structure of a protein molecule. The mutation itself represents a highly localized change in one of the nucleotide pairs in the chromosomal DNA molecule (see fig. 3.4).

Since persons with sickle-cell anemia ordinarily do not survive to reproductive age, it might be expected that the detrimental gene would pass rapidly from existence. Each failure of the homozygous anemic individual to transmit his genes would result each time in the loss of two aberrant genes

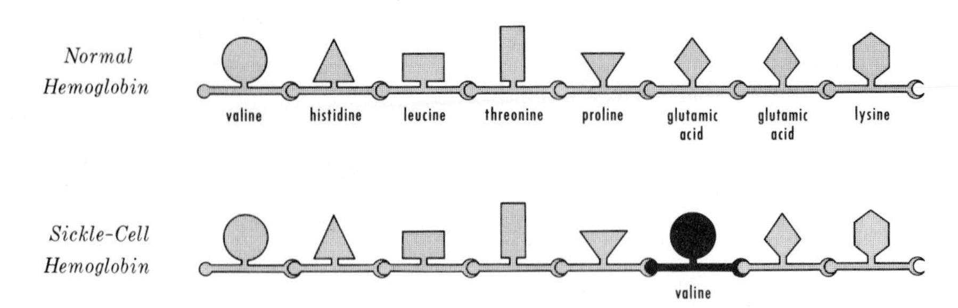

Figure 7.4 Amino acid sequences in a small section of the normal hemoglobin molecule and of the sickle-cell hemoglobin. The substitution of a single amino acid, glutamic acid by valine, is responsible for the abnormal sickling of human red blood cells. (Based on studies by Vernon Ingram.)

from the population. And yet, the sickle-cell gene reaches remarkably high frequencies in the tropical zone of Africa. In several African populations, 20 percent or more of the individuals have the sickle-cell trait, and frequencies as high as 40 percent have been reported for some African tribes. The sickle-cell trait is not confined to the African continent; it has been found in Sicily and Greece, and in parts of the Near East. What can account for the high incidence of the sickle-cell gene, particularly in light of its detrimental action?

Theory of Balanced Polymorphism

For simplicity in the presentation of population dynamics, we will consider the sickling gene to be recessive. Accordingly, we can symbolize the normal gene as *A* and its allele for sickling as *a*. Now, the explanation for the high level of the deleterious sickle-cell gene is to be found in the possibility that the heterozygote (*Aa*) is superior in fitness to *both* homozygotes (*AA* and *aa*). In other words, selection favors the heterozygote and both types of homozygotes are relatively at a disadvantage. Let us examine the theory behind this form of selection.

Figure 7.5 illustrates the theory. The classical case of selection discussed in preceding chapters is portrayed in the first part of the figure. In this case, when *AA* and *Aa* individuals are equal in reproductive fitness, and the *aa* genotype is completely selected against, the recessive gene will be eliminated. Barring recurring mutations, only *A* genes will ultimately be present in the population.

Let us assume in another situation (case II in fig. 7.5) that both homozygotes (*AA* as well as *aa*) are incapable of leaving surviving progeny. The only effective members in the population are the heterozygotes (*Aa*). Obviously, the frequency of each gene, *A* and *a*, will remain at a constant level of 50 percent. The inviability of both homozygous types probably never exists in nature, but the scheme does reveal how a stable relation of two

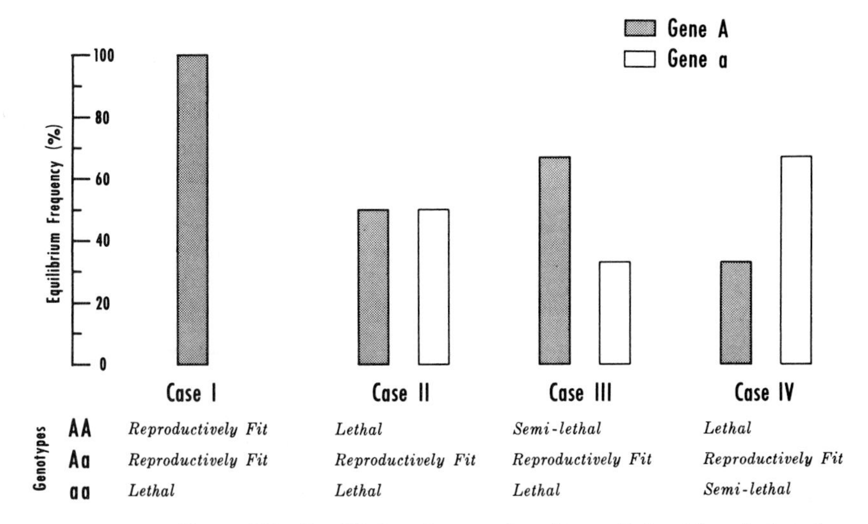

Figure 7.5 Equilibrium frequencies of two alleles (*A* and *a*) under different conditions of selection (ignoring mutation). In contrast to case I (complete selection and total elimination of *a*), the recessive gene can be retained at appreciable frequencies in a population when the heterozygote (*Aa*) is superior in reproductive fitness to *both* homozygotes (cases II, III, and IV, showing different relative fitnesses of the two homozygotes, *AA* and *aa*).

alleles at high frequencies is possible. If, as in case III, the *AA* genotype leaves only half as many progeny as the heterozygote, and the recessive homozygote is once again inviable, then it is apparent that more *A* alleles are transmitted each generation than *a* alleles. Eventually, however, the *A* gene will reach an equilibrium point at 0.67. Here, it may be noted that, although the recessive homozygote is lethal, the frequency of the recessive allele (*a*) is maintained at 0.33. In the last illustrative example (case IV), the recessive homozygote is not as disadvantageous as the dominant homozygote, but both are less reproductively fit than the heterozygote. Here also, both genes remain at relatively high frequencies in the population. Indeed, the recessive allele (*a*) will constitute 67 percent of the gene pool.

We have thus illustrated in simplified form the selective forces that serve to maintain two alleles at appreciable frequencies in a population. This phenomenon is known as *balanced polymorphism.* The loss of a deleterious recessive gene through deaths of the homozygotes is balanced by the gain resulting from the larger numbers of offspring produced by the favored heterozygotes.

Experimental Verification of Balanced Polymorphism

A simple experiment, using the fruit fly, *Drosophila melanogaster,* can be performed to demonstrate that an obviously lethal gene can be maintained at a stable, relatively high frequency in a population. The geneticist P. M. Sheppard of the University of Liverpool in England introduced into a breeding cage a population of flies of which 86 percent were normal and 14 percent carried the mutant gene *stubble,* which affects the bristles of the fly. The stubble gene is lethal when homozygous; hence, all stubble individuals are heterozygous. Ordinarily the heterozygous fly does not have any reproductive advantage over the normal homozygote. Sheppard, however, created a situation whereby the heterozygote would be favored by removing 60 percent of the normal flies from the population each generation. Consequently, the heterozygote was rendered superior in fitness to either homozygote by virtue of the natural lethality of the stubble homozygote and the enforced reproductive incapacity of many of the normal homozygotes. The results of the experiment are recorded in table 7.1.

Table 7.1*
Experimental Balanced Polymorphism

Generation	Percentage of Stubble Flies	Frequency of Stubble Gene
1	14.3	0.0715
2	33.7	0.1685
3	57.6	0.2880
4	63.2	0.3160
5	69.1	0.3455
6	73.5	0.3675
7	72.9	0.3645
8	73.4	0.3670
9	72.9	0.3645

* From P. M. Sheppard, *Natural Selection and Heredity*. New York: Philosophical Library, 1959.

It should be noted that the frequency of the stubble gene increased in the early generations, but then became stabilized at a level of about 0.365. The equilibrium level is reached when as many stubble genes are lost from the population, through death of the stubble homozygotes, as are gained as a result of the reproductive advantage of the heterozygote. Although the stubble gene is lethal, the population, under the constant conditions of the experiment, remained at a stable state with a high number of heterozygotes.

The frequency of the stubble gene will fall rapidly when the usual reproductive potential of the normal homozygote is restored. When the normal homozygotes and heterozygotes are equal in reproductive fitness (refer to case I, fig. 7.5), the normal gene will supplant the abnormal stubble gene in the population. With these considerations in mind, we shall return to our discussion of sickle-cell anemia.

Superiority of the Heterozygotes

The high frequency of the sickle-cell gene in certain African populations can be explained by assuming that the heterozygotes (individuals with sickle-cell trait) have a selective advantage over the normal homozygotes. What might be the nature of the advantage? Field work undertaken in Africa in 1949 by the British geneticist Anthony Allison revealed that the incidence of the sickle-cell trait is high in regions where malignant subtertian malaria caused by the parasite *Plasmodium falciparum* is hyperendemic—that is, transmission of the infection occurs throughout most of the year. Thus, the population is almost constantly reinfected with malaria, and susceptible individuals suffer from severe and prolonged attacks of malaria. Under such circumstances, relative immunity to falciparum malaria would be most beneficial.

Allison examined blood from African children and found that carriers of the sickle-cell trait were relatively resistant to infection with *Plasmodium falciparum*. The heterozygous carriers were infected less often with the parasite than the homozygous dominant nonsicklers. Moreover, among those heterozygotes that were infected, the incidence of severe, or fatal, attacks of malaria was strikingly low. The evidence is strong that the sickle-cell gene affords some degree of protection for young children against malarial infection. Hence, in areas where malaria is common, children possessing the sickle-cell trait will tend to survive more often than those without the trait, and are more likely to pass on their genes to the next generation. The heterozygotes (*Aa*) are thus superior in fitness to both homozygotes, which are likely to succumb either from anemia on the one hand (*aa*) or malaria on the other (*AA*).

Frank B. Livingstone of the University of Michigan has impressively revealed the relationship among the spread of agriculture, malaria, and sickle-cell anemia. The spread of the sickling gene was greatly enhanced by the development of agriculture. The clearing of the forest in the preparation of ground for cultivation provided new breeding areas for the mosquito *Anopheles gambiae*, the vector, or carrier, of the malaria plasmodium. The spread of malaria has been responsible for the spread of the selective advantage of the sickle-cell gene, which in the heterozygous state imparts resistance to malaria. In essence, the selective advantage of the heterozygote tends to increase in direct proportion to the amount of malaria present in a given area. Hunting populations in Africa show a very low incidence of malaria and an equally low frequency of the sickling gene. The Pygmies of the Ituri Forest constitute a good example.

Communities in Africa with the greatest reliance on agriculture (rather than on hunting or animal husbandry) tend to have the highest frequencies of the sickle-cell trait. A high incidence of the sickle-cell trait in an intensely malarious environment has the consequence of reducing the number of individuals capable of being infected by the malarial parasite and, accordingly, of lowering the mortality from such infections. More human energy, or greater manpower, is thus made available for raising and harvesting crops. Ironically, then, the sickle-cell trait carries with it the beneficent effect of enabling tribes to develop and maintain an agricultural culture rather than adhere to a hunting or pastoral existence. This is a curious, but striking, instance of the interplay of biological change and socioeconomic adaptation.

In a similar finding, evidence exists that persons whose red blood cells are deficient in the enzyme G6PD (glucose-6-phosphate-dehydrogenase) are less likely to be affected by malaria. Data also show a strong correlation between the incidence of malaria and the frequency of thalassemia (a type of anemia) in the Italian peninsula and in Sardinia. Malaria apparently has had a profound influence on human events.

Disruption of Balanced Polymorphism

We should expect the frequency of the sickle-cell gene to be low in malaria-free areas, where the selective advantage of the heterozygote would be removed. Accordingly, we find that the lowest frequencies of the sickle-cell gene occur consistently in regions relatively free of malaria. The frequency of the sickle-cell gene has fallen to relatively low levels in the Negro population of the United States. The frequency of the sickle-cell trait among American blacks is currently about 9 percent. The incidence at birth of the disabling sickle-cell anemia is estimated at 2 per 1,000 individuals.

The disappearance of malaria disrupts the balanced polymorphic state, and the gene for sickle-cell anemia begins to decline in frequency. If the reproductive fitness of the heterozygous carrier is equal to that of the normal homozygote, and the fitness of the sickle-cell anemic individual is less than normal, then the incidence of the heterozygous carrier will become depressed generation after generation. At present, medical researchers are attempting to find a cure for sickle-cell anemia. If sickle-cell anemic individuals could be completely cured, then the selective process would be thwarted and the sickle-cell gene would no longer decline in frequency.

Tay-Sachs Disease

Tay-Sachs disease is a recessively inherited condition that is untreatable and fatal. The basic defect is the abnormal accumulation in the brain cells of a fatty substance (specifically, ganglioside GM_2) as a consequence of the absence of a particular enzyme (hexosaminidase A). The storage of massive amounts of the lipid leads to profound mental and motor deterioration. Affected children appear normal and healthy at birth, but within six months the nerves of the brain and spinal cord exhibit marked signs of deterioration. Listless and irritable at first, the infant finds it increasingly difficult to sit up or stand. By the age of one year, the child lies helplessly in his crib. He becomes mentally retarded, progressively blind, and finally paralyzed. The disease takes its lethal toll by the age of three to four years. There are no known survivors and no cure.

A feature of special interest is that 9 out of 10 affected children are of Jewish heritage. It is especially common among the Ashkenazi Jews of northeastern European origin, particularly from provinces in Lithuania and Poland. In the United States, Tay-Sachs disease is about 100 times more prevalent among the Ashkenazi Jews than among other Jewish (Sephardi) groups and non-Jewish populations. It is estimated that 1 of 40 Jewish persons is heterozygous, whereas only 1 of 380 non-Jewish persons is a heterozygous carrier. If the high incidence of heterozygotes is maintained by mutation alone, then an extraordinarily high mutation rate of the detrimental recessive gene would have to be postulated. However, some reproductive advantage for the heterozygote carrier would seem the most plausible explanation.

It has been suggested that the Ashkenazi Jews, who have lived for several millennia in the urban ghettos in Poland and the Baltic states, have been exposed to different selective pressures than other Jewish groups who have lived in countries around the Mediterranean and Near East. The densely populated urban ghettos may have experienced repeated outbreaks of infectious diseases. In 1972, the geneticist Ntinos Myrianthopoulos of the National Institutes of Health in Bethesda, Maryland, presented data that show that pulmonary tuberculosis is virtually absent among grandparents of children afflicted with Tay-Sachs disease, although the incidence of Jewish tuberculosis patients from eastern Europe is relatively high. The findings indicate that the heterozygous carrier of Tay-Sachs disease is resistant to pulmonary tuberculosis in regions where this contagious disease is prevalent.

Cystic Fibrosis

One of the most serious recessive disorders among children is *cystic fibrosis* of the pancreas. About 6,000 infants are born with the disease each year in the United States. There are at least 5 million heterozygous carriers, and 1 child in 2,500 is born with the disease. Virtually all patients with cystic fibrosis are Caucasians; Negroes and Orientals are rarely afflicted.

Less than 35 years ago, cystic fibrosis was not even recognized as a distinct medical entity. Today it has the unenviable reputation of being one of the most important disorders of childhood. Intestinal obstruction is the first symptom of the disease; the infant's stools are frequent, large, and foul. The pancreas secretes an abnormally thick or viscid mucous material that blocks digestion in the intestinal tract. In addition, all mucus-secreting tissues of affected infants are abnormal. The sticky mucus produced by the lungs is particularly serious; the lungs become clogged and the child has repeated bouts of pneumonia. A characteristic finding, by which cystic fibrosis is readily diagnosed, is the increased amounts of salt in the sweat.

Before the introduction of antibiotics, affected children invariably died in infancy of constant infection. New drugs, inhaled as vapor, soften the thick mucus of the lungs and have enabled some children to weather the difficult first years. The widespread application of antibiotics has led to a considerable reduction in early mortality. In fact, several women patients with cystic fibrosis responded so well to the special treatment that they now have children of their own. All these children are phenotypically normal. But these normal children born of affected mothers harbor the recessive gene for cystic fibrosis and can pass on the defective gene to their offspring.

A reproductive advantage for the heterozygote would account for the high frequency of the recessive gene for cystic fibrosis in Caucasian populations. The magnitude of the selective advantage of the heterozygous carrier over the normal homozygote need only be 2 percent to maintain an incidence of the disorder at 1 in 2,500 births. To ascertain the reproductive advantage, one could obtain information on the fertility of heterozygous parents. However, the overall reproductive performance of heterozygous parents tends to be biased by the birth of a child with cystic fibrosis and by the knowledge that this lethal hereditary disorder can arise in subsequent pregnancies.

An unbiased estimate of the fertility of unsuspecting heterozygotes, who unknowingly perpetuate the gene, can be obtained by assessing the fertility of the *grandparents* of the affected children. It may be assumed that at least one of the grandparents (maternal or paternal) of an affected child is a heterozygote. Thus, the size of the sibship from grandparental matings of the type *AA* x *Aa* (where *a* is the allele for cystic fibrosis) can be compared with the sibship size of grandparental matings of the type *AA* x *AA*. It should be noted that family size cannot be attributed to knowledge of the disorder since by the time a grandchild with cystic fibrosis has been identified, the fertility of the grandparents has been completed. Such a study was undertaken in 1967 by Alfred Knudson and his co-workers at the City of Hope Medical Center at Duarte, California. The grandparents of patients with cystic fibrosis were found to have an average of 4.34 offspring compared with 3.43 offspring for the appropriate control group. Thus, on the average, matings of type *AA* x *Aa* have one additional child compared with *AA* x *AA* matings. The greater reproductive fitness of the heterozygous carrier suggests that the detrimental gene for cystic fibrosis is being maintained at exceptionally high frequencies by balanced polymorphism.

Implications of Balanced Polymorphism

According to the classical concept of selection (discussed in chapter 5), a deleterious gene has a harmful effect when homozygous and virtually no expression in the heterozygous state. Deleterious genes will be reduced in frequency by selection to very low levels, and will be maintained in the population by recurrent mutation. The fittest individuals are homozygous for the normal, or "wild-type," allele at most loci. Stated another way, the fittest individuals carry relatively few deleterious genes in the heterozygous state.

In this chapter, we have considered examples of genes that impair fitness when homozygous and actually improve fitness in the heterozygous state. To what extent are individuals heterozygous at their gene loci? Geneticists estimate that 20 to 50 percent of the loci in an individual exist in two or more allelic forms. If the alternative alleles at these polymorphic loci are maintained by selection that favors the heterozygote, then low fitness would have to be assigned to an unusually large number of homozygous loci. In other words, an unreasonably high level of unfitness would prevail in the population because of the selective disadvantage of many alleles in the homozygous state.

The possibility exists that most of the alternative alleles at polymorphic loci are *selectively neutral*—that is, the different alleles at one locus confer neither selective advantage or disadvantage on the individual. Much of the observed variation, then, would represent merely the accumulation of neutral mutations. The notion of selectively neutral alleles has been much debated. Are there mutant alleles whose effects on fitness are not at all different from the more frequent alleles that lead to a normal phenotype? And, can some of these neutral mutant alleles reach frequencies purely by chance as high as the frequencies that characterize the state of balanced polymorphism? These vexing questions have yet to be resolved satisfactorily.

Selected Readings

Allison, A. C. 1956. Sickle cells and evolution. *Scientific American,* August, pp. 87–94.

Anfinsen, C. B. 1963. *The molecular basis of evolution.* New York: John C. Wiley & Sons.

Chernoff, A. 1959. The distribution of the thalassemia gene: a historical review. *Blood* 14:899–912.

Dobzhansky, T. 1962. *Mankind evolving.* New Haven: Yale University Press.

Harris, H. 1969. Enzyme and protein polymorphism in human populations. *British Medical Bulletin* 25:5–13.

Ingram, V. 1963. *The hemoglobins in genetics and evolution.* New York: Columbia University Press.

Knudson, A. G., Wayne, L., and Hallett, W. Y. 1967. On the selective advantage of cystic fibrosis heterozygotes. *American Journal of Human Genetics* 19:388–392.

Livingstone, F. B. 1958. Anthropological implications of sickle cell gene distribution in West Africa. *American Anthropologist* 60:533–62.

Livingstone, F. B. 1967. *Abnormal hemoglobins in human populations.* Chicago: Aldine Publishing Co.

Myrianthopoulos, N. C., and Aronson, S. M. 1966. Population dynamics of Tay-Sachs disease. I. Reproductive fitness and selection. *American Journal of Human Genetics* 18:313–327.

Myrianthopoulos, N. C. 1972. Founder effect in Tay-Sachs disease unlikely. *American Journal of Human Genetics* 24:341–342.

Price, J. 1967. Human polymorphism. *Journal of Medical Genetics* 4:44–67.

Sheppard, P.M. 1959. *Natural selection and heredity.* New York: Philosophical Library.

Wiesenfeld, S. L. 1967. Sickle-cell trait in human biological and cultural evolution. *Science* 157:1134–40.

8

Selection and Human Blood Antigens

The mother-child blood incompatibility in humans, known as *erythroblastosis fetalis,* represents an interesting case of selection *against* the heterozygote. The erythroblastotic infant is always the heterozygote (Rr), born to Rh-negative mothers and Rh-positive fathers. Each death of an erythroblastotic infant (Rr) results in the elimination of one R and one r gene. Given a population in which the two genes are unequal in frequency, continual selection against the heterozygote would result in a greater relative loss of the less frequent gene. Eventually, the rarer of the two genes would become lost, or decline to a low level to be maintained solely by mutation.

In populations where the R gene is much more common than the r allele, we should be witnessing a gradual dwindling of the r gene. No decline, however, in the frequency of the r gene is evident in most Caucasian populations. One counterbalancing factor might be the tendency of parents who have lost infants from erythroblastosis to compensate for their losses by having relatively large numbers of children. Thus, if a father is heterozygous (Rr) and the mother is homozygous (rr), there is an even chance that the infant will be rr and unaffected. Each unaffected child born restores two r genes lost by the death of two Rr erythroblastotic sibs. Accordingly, an excess of homozygous children (rr) counterbalances the r genes lost through erythroblastosis. This consideration alone would override the selective force against the heterozygote. However, careful analyses of the available data have failed to support the reality of "reproductive compensation."

We shall examine the circumstances that might account for the persistence of the r gene despite strong selection against the heterozygote, and consider the intricate interaction between Rh incompatibility and ABO incompatibility.

Rh Incompatibility

The discovery of the Rh factor in the 1940s permitted investigators to deduce the real nature of a relatively common blood disease of the newborn. Affected newborn infants are burdened with anemia, jaundice, enlargement of the liver and spleen, and heart failure. The condition is known variously as *erythroblastosis fetalis, hemolytic disease of the newborn,* or simply *Rh disease.* The mother of an afflicted baby lacks a component in her red blood cells—a component called the Rh factor (antigen) because it was first isolated in the rhesus monkey. She is said to be "Rh-negative." In contrast, the father and the affected child possess the Rh factor (antigen) and both are "Rh-positive."

The Rh antigen in the blood cell is controlled by a dominant gene, designated *R*. An Rh-positive person has the dominant gene, either in the homozygous (*RR*) or heterozygous (*Rr*) state. All Rh-negative individuals carry two recessive genes (*rr*) and are incapable of producing the Rh antigen. The inheritance of the Rh antigen follows simple Mendelian laws. A mother who is Rh-negative (*rr*) need not fear having Rh-diseased offspring if her husband is likewise Rh-negative (*rr*). If the husband is heterozygous (*Rr*), half of the offspring will be Rh-negative (*rr*) and none of these will be afflicted. The other half will be Rh-positive (*Rr*), just like the father, and are potential victims of hemolytic disease. If the Rh-positive father is homozygous (*RR*), then all the children will be Rh-positive (*Rr*) and potential victims. In essence, an Rh-positive child carried by an Rh-negative mother is the setting for possible, though not inevitable, trouble.

The chain of events leading to hemolytic disease begins with the inheritance by the fetus of the dominant *R* gene of the father. Rh antigens are produced in the red blood cells of the fetus. The fetal red cells bearing the Rh antigens escape through the placental barrier into the mother's circulation, and stimulate the production of antibodies (anti-R) against the Rh antigens on the fetal red cells (fig. 8.1). The mother, having produced antibodies, is said to be immunized (or sensitized) against her baby's blood cells. The maternal antibodies almost never attain a sufficient concentration during the first pregnancy to harm the fetus. In fact, although fetal cells cross the placenta throughout pregnancy, they enter the maternal circulation in much larger numbers during delivery, when the placental vessels rupture. It is now generally conceded that sensitization of the mother takes place shortly *after* the delivery of the first Rh-positive child. Accordingly, the first-born is rarely affected, unless the mother had previously developed antibodies from having been transfused with Rh-positive blood or has had a prior pregnancy terminating in an abortion.

The antibodies remain in the mother's system, and may linger for many months or years (fig. 8.1). If the second baby is also Rh-positive, the mother may send sufficient antibodies into the fetus' bloodstream to destroy its red

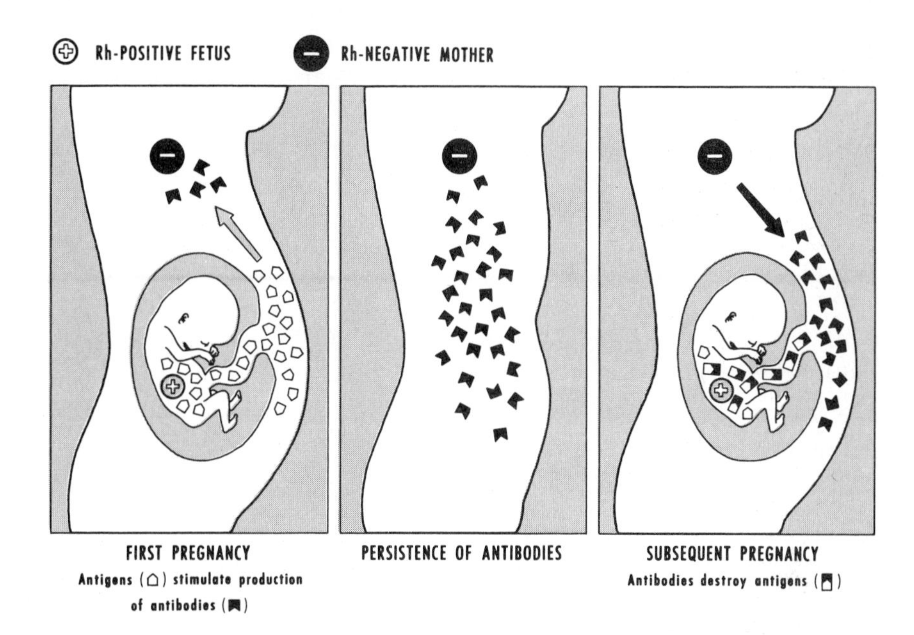

Figure 8.1 Rh disease in the newborn. The first baby is rarely harmed, but subsequent babies are potential victims of the mother's antibodies which are capable of destroying the infant's blood cells.

cells. The majority of affected fetuses survive for the usual gestation period but are born in critical condition from anemia. Severely anemic individuals are likely to be jaundiced—that is, the skin assumes a deep yellow-orange tint. When the fetus' red cells hemolyze (break down), the hemoglobin liberated from the ruptured cells is transformed into a yellow pigment called *bilirubin*. The liver ordinarily would convert the bilirubin into harmless bile. But the amount of yellow bilirubin in the infant is unmanageable and accumulates in the blood, turning the plasma into an almost yellow-orange liquid. The infant is consequently jaundiced.

Excessive amounts of bilirubin pose a grave threat to the newborn infant. During pregnancy fetal bilirubin is transported across the placenta and eliminated by the mother. From the time of birth, however, bilirubin accumulates as the affected infant fails to dispose of it. Bilirubin has been shown to be highly toxic to the soft brain tissues; the brain may be permanently damaged.

Another aspect of the hemolytic condition, which has given it its original name (erythroblastosis fetalis), is the presence of immature red cells (the erythroblasts) in the circulating blood. It is as if the liver and the spleen in an attempt to combat the severe anemic condition produce vast numbers of "unfinished" red blood cells. Unattended, erythroblastosis fetalis leads to stillborn or neonatal death. Many of the erythroblastotic babies are saved by exchange transfusion of Rh-negative blood; others, however, die despite treatment. Exchange transfusion is essentially a flushing-out process, whereby the infant's blood is gradually diluted with Rh-negative donated blood until at the end of the procedure most of the infant's circulating blood is problem-free. In severe cases, where it has been predicted that the fetus would die before it was mature enough for premature delivery, intrauterine transfusions have been used successfully.

As might have been anticipated, the Rh gene has turned out to be more complex than initially envisioned. There are several variant alleles, and a corresponding diversity of antigenic constitutions. This diversity need not concern us here. The most common antigen is the one that was first recognized, known more specifically now as Rh_0 or D. It is the presence of Rh_0 that is tested in ordinary clinical work.

Incidence of Rh Disease

Among Caucasians in the United States, the incidence of Rh-negative persons is approximately 16 percent. In certain European groups, such as the Basques in Spain, the frequency of Rh-negative individuals rises as high as 34 percent. Non-Caucasian populations are relatively free of Rh hemolytic disease. The incidence of Rh-negative persons among the full-blooded American Indians, Eskimos, African Negroes, Japanese, and Chinese is one percent or less. In contrast, the frequency of Rh-negative American Negroes is high (9 percent).

We can calculate the frequencies of marriages that contain risk with respect to Rh hemolytic disease. Table 8.1 shows the different types of marriages that can occur when individuals choose their mates by sheer chance—that is, without regard to each other's Rh makeup. Of the two kinds of marriages that can result in Rh-positive pregnancies, 5.76 percent would be *rr* (females) x *RR* (males) and 7.68 percent would be *rr* (females)

Table 8.1

Frequencies of Marriages by Chance of Rh-positive and Rh-negative Individuals in Caucasian Populations*

	Male (♂)		
Female (♀)	36% *RR*	48% *Rr*	16% *rr*
36% *RR*	12.96% *RR* x *RR*	17.28% *RR* x *Rr*	5.76% *RR* x *rr*
48% *Rr*	17.28% *Rr* x *RR*	23.04% *Rr* x *Rr*	7.68% *Rr* x *rr*
16% *rr*	5.76% *rr* x *RR*	7.68% *rr* x *Rr*	2.56% *rr* x *rr*

* Rh-negative persons make up 16% of the population; among Rh-positive persons, 36% are homozygous dominant and 48% are heterozygous.

x *Rr* (males). Thus, approximately 13 percent of all marriages, or 1 out of every 8, carry an erythroblastosis fetalis risk.

From the above data, we can ascertain the frequency of potentially dangerous pregnancies. All pregnancies from the *rr* (females) x *RR* (males) would yield an Rh-positive fetus, but only one half of the infants from the *rr* (females) x *Rr* (males) would be Rh-positive. Accordingly, the frequency of all potentially troublesome pregnancies is $5.76 + 3.84 = 9.60$ percent, or slightly less than one-tenth of all pregnancies.

Theoretically, then, 1 out of 9 or 10 pregnancies should result in an erythroblastotic child. However, in actuality, only 1 in 200 pregnancies results in an affected baby. The reasons for the low observed incidence are not entirely clear. Apparently, differences in the ease of sensitization exist among Rh-negative women. The Rh antigen of the fetus may fail to get through the placenta, or some Rh-negative mothers are incapable of responding to the antigen. The low occurrence may also reflect a peculiar protective role of the ABO blood groups in reducing the risk of Rh incompatibility. This curious interaction requires an understanding of the ABO blood system.

ABO Blood Groups

The Rh factor is not the only antigen that a red blood cell may possess on its surface. Red blood cells may also carry two other major antigens, called A and B. The A and B factors were discovered in the early 1900s by the Austrian pathologist Karl Landsteiner. His discovery led to the identification of the four familiar major human blood groups or types—A, B, AB, and O.

A person may have one or the other of the two antigens (A or B) in his blood cells, or he may have both or neither. A person possessing antigen A is said to belong to blood group *A*; a person having antigen B belongs to group *B*; a person having both antigens is in group *AB*; and a person having neither belongs to group *O*. Landsteiner showed that there are two corresponding antibodies in the serum of the blood, which he designated as anti-A and anti-B. If serum containing anti-A is mixed with a suspension of blood cells bearing antigen A, the cells will clump together in large granular masses. On the other hand, if anti-A is mixed with B blood cells, or anti-B with A blood cells, there is no reaction; the cells remain suspended without clumping. This simple principle is the basis for blood-group tests.

Whatever antigen a person has in his cells, the corresponding antibody is lacking in the serum (table 8.2); a person obviously does not have the antibody that can destroy his own red cells. He may, however, have the antibody that reacts against the antigen possessed by another person. For example, type A individuals lack anti-A but contain the antibody against B. This knowledge provides the basis for the successful transfusion of blood between individuals. The cardinal rule is to avoid introducing antigens that can be destroyed (or agglutinated) by antibodies in the serum of the recip-

Table 8.2

ABO Blood Groups in Man

Group	Genotype	Antigens in Cells	Antibodies in Serum	Can Donate Blood to:	Can Receive Blood from:
A	I^AI^A or I^Ai	A	anti-B	A, AB	A, O
B	I^BI^B or I^Bi	B	anti-A	B, AB	B, O
AB	I^AI^B	A & B	none	AB	A, B, AB, O
O	ii	none	anti-A & anti-B	A, B, AB, O	O

ient. Antibodies in the donor are inconsequential because the amount of blood transfused is small relative to the total blood volume of the recipient, and the antibodies are quickly diluted out by the recipient's blood. Type O persons are the acknowledged "universal donors" since they contain no antigens that could be acted on by the recipient's antibodies. Type AB persons lack both antibodies and, accordingly, can receive blood from persons of all types without fear of destroying the cells contributed by the donor.

Inheritance of the ABO Blood Groups

In humans, the hereditary basis of a trait is sought by observing the numbers and distributions of the phenotypic variants in families. The blood groups serve as an excellent example of how the mode of inheritance has been deduced from an analysis of large numbers of family pedigrees. Representative marriages are shown in figure 8.2.

When both parents are AB, only three types appear among the offspring: A, B, and AB (cross 1 in fig. 8.2). Type O children do not arise from such marriages, and the AB offspring arise twice as often as either the A or B sib. The ¼ A: ½ AB: ¼ B phenotypic ratio is clearly the result to be expected from the mating of two heterozygous parents. Each heterozygous parent can be said to carry an allele (designated I^A) responsible for the production of antigen A as well as an allele (I^B) responsible for antigen B. Both alleles are denoted by capital letters, since neither one is dominant to the other and both express themselves in the heterozygote.

Cross 2 in figure 8.2 reveals that type O parents have only O children. This suggests that the parents and their offspring are homozygous for a recessive gene (designated i). The relation between the three alleles (I^A, I^B, and i) becomes clear from the analysis of AB x O marriages (cross 3, fig. 8.2). Children from such marriages are either A or B, in equal proportions, and never O. Thus, genes I^A and I^B, although exhibiting no dominance with respect to each other, are each dominant to allele i.

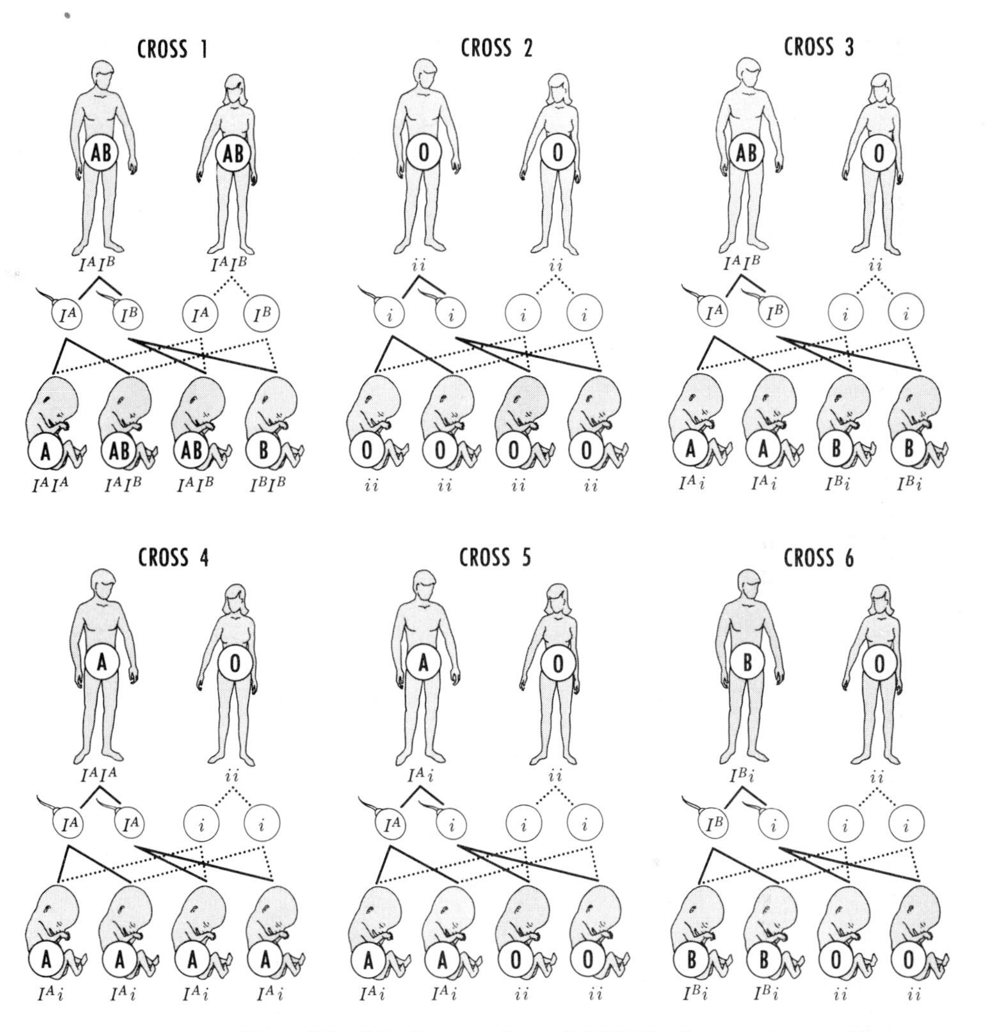

Figure 8.2 Inheritance pattern of ABO blood groups in man. Three different genes are involved (I^A, I^B, and i), but a given person can possess only two of the three genes.

The outcome of an A x O marriage depends on whether the A parent is homozygous or heterozygous (crosses 4 and 5, fig. 8.2). If homozygous (I^AI^A), only A children are possible; if heterozygous (I^Ai), both A and O children can be produced. Likewise, as seen in cross 6 (fig. 8.2), type O children can arise from a B x O marriage only when the B parent is heterozygous (I^Bi).

We can thus recognize that there are three alternative forms, or three alleles, of one gene: (1) I^A for production of the A antigen; (2) I^B for the production of the B antigen; and (3) i, which results in neither A nor B antigen. The i allele is recessive to both I^A and I^B, but I^A and I^B are codominant. *Only two of the three alleles may normally occur in an individual and only one in the gamete.*

We assume that at some time in the past mutations occurred at this one gene locus. However, we do not know which of the three is the original allele, and which two are the mutant alleles. Further analysis has revealed two major subdivisions of group A, based on the presence of the red cells of two separate antigenic factors, A_1 and A_2. There are accordingly more than three alleles at this one locus. Since a given gene contains many sites at which mutations can occur, it is not surprising that different alleles with dissimilar functions have arisen.

ABO Incompatibility

It may happen that the fetus contains an A or B antigen (inherited from the father) that is not present in the mother. For example, she may be type O, and her baby may be type A. The mother's serum in this case would carry the naturally occurring antibodies—anti-A and anti-B. The anti-A of the mother may enter the fetal circulation through the placenta and destroy the fetus' red cells carrying the A antigen. This may result in hemolytic disease of the newborn, in which the infant is delivered with anemia and jaundice. Hemolytic disease due to anti-A (or anti-B) is much milder than that caused by the familiar anti-Rh, and rarely requires an exchange transfusion of blood. Hemolytic disease due to ABO incompatibility, however, often occurs in the firstborn. This is quite different from the pattern of Rh disease.

Theoretically, 20 percent of all babies have ABO blood types potentially incompatible with the mother's type. The actual number of clinically affected infants is below expectation; about 1 in 150 of all births. Evidently, many newborn of ABO incompatible pregnancies are protected in some way. An examination of table 8.3 reveals that type AB women never have "incompatible" babies, and type O babies are always "compatible." Clinically, it has been observed that if a baby has hemolytic disease due to anti-A or anti-B, the mother is usually type O, and very seldom type A or B. Type O

Table 8.3

ABO Groups of Mothers and Their Babies

Mother's Type	Antibodies in Mother's Serum	Types of Incompatible Babies	Types of Compatible Babies
A	anti-B	B, AB	O, A
B	anti-A	A, AB	O, B
AB	none	none	A, B, AB
O	anti-A & anti-B	A, B	O

women have appreciable quantities of antibodies of low molecular weight (the so-called 7S gamma globulin fraction), which readily cross the placental barrier.

A disquieting finding in recent years is that anti-A and anti-B manifest their harmful effects primarily in early pregnancy. There is a significantly greater rate of spontaneous abortion among type O women married to men of type A or B than among A or B women married to O men. The fetal loss is an estimated 20 percent of all A or B conceptuses from marriages of an A or B father and an O mother, as compared in the reciprocal marriage. In terms of large numbers of pregnancies terminating prematurely, ABO incompatibility looms as more frightening than Rh incompatibility. Ironically, as we shall see below, ABO incompatibility between mother and fetus may have beneficial consequences under certain circumstances.

Interaction of ABO and Rh Incompatibility

In 1943, Philip Levine called attention to the fact that infants of Rh-negative, type O mothers develop Rh disease less often than those of type A or B, Rh-negative mothers. If the infant's red cells are type A or type B, the maternal anti-A and anti-B antibodies in type O women destroy the infant's Rh-positive red cells when they enter the Rh-negative mother's bloodstream. Thus, the invading fetal cells are eliminated before they have the opportunity to sensitize the mother. Strange as it may seem, then, the ABO incompatibility protects the mother and child against a simultaneous Rh incompatibility. A double incompatibility clearly affords a lower risk of fetal loss than Rh incompatibility alone. The general picture is that immunization to the Rh antigen, with rare exceptions, occurs only when the ABO blood type of the fetus is compatible with that of the mother.

The Control of Rh Disease

In the 1960s, Vincent Freda and John Gorman at the Columbia-Presbyterian Medical Center in New York and William Pollack at the Ortho Research Foundation in New Jersey sought the means to suppress the production of antibodies in mothers who had recently delivered an Rh-positive infant. Experiments performed 50 years earlier by the distinguished American bacteriologist Theobald Smith furnished an important clue to the solution. In 1909, Smith arrived at the general principle that passive immunity can prevent active immunity. That is, an antibody given passively by injection can inhibit the recipient from producing its own active antibody. After five years of experimentation and testing, Freda, Gorman, and Pollack successfully developed an immunosuppressant consisting of a blood fraction (gamma globulin) rich in Rh antibodies. Injected into the bloodstream of the Rh-negative mother no later than three days after the birth of her first Rh-positive child, the globulin-Rh antibody preparation (known as RhoGAM) suppresses the mother's antibody-making activity. Several hundred mothers have already received the preparation; none have formed active antibodies. More impressively, some of them have now delivered a second Rh-positive baby, and none of the babies were afflicted with Rh disease. The evidence is overwhelming that the immunosuppressant is effective.

The immunosuppressant does not prevent hemolytic disease if maternal antibody is already present by earlier pregnancies or by previous transfusion of Rh-positive blood. For this reason the preparation is administered only to Rh-negative women lacking anti-Rh in their sera at the time of delivery. With each delivery, opportunity for exposure to Rh-positive fetal cells is repeated. The protection given at the delivery of the first infant does not protect the mother from exposure to antigen received at a subsequent pregnancy. Here, the immunosuppressant must be given immediately following each pregnancy.

We commented earlier on the relatively high incidence of Rh-negative persons among Caucasians (16 percent). The British geneticist J. B. S. Haldane suggested that the present gene frequencies among Europeans are probably the result of comparatively recent admixtures (about 10,000 years ago) of populations that were wholly or predominately *RR* with populations that were then largely *rr*. This would establish communities that would be highly polymorphic for *R* and *r*. Haldane calculated that 600 generations, or 15,000 years, of continual selection against heterozygotes would be required to reduce the frequency of Rh-negative individuals from the present 16 percent to 1 percent. Haldane's calculations were premised on the absence of medical therapy to save the lives of the heterozygotes. With the advent of RhoGAM, we may forsee not only the end of Rh disease as a major clinical problem but also the abridgment of natural selection against the heterozygote. We may anticipate that the recessive *r* allele will not decline in the human gene pool, but rather endure.

Blood Groups and Disease

Suggestive evidence exists that individuals belonging to different ABO groups differ in their susceptibility to certain common diseases. Type O persons are more prone to develop peptic ulcers than individuals belonging to groups A, B, or AB. On the other hand, type O individuals are less likely to develop pernicious anemia or cancer of the stomach than type A persons.

The association between diseases and the ABO blood groups is difficult to comprehend or evaluate. Some diseases, such as cancer of the stomach, mainly affect a person beyond the reproductive age, after the person has passed on his genes to his descendants. Thus, we would not expect the frequencies of the blood group genes to change appreciably in successive generations. Recall that selection operates only when an individual is so affected that, compared with those of other genotypes, he cannot contribute proportionately to the population's gene content in the following generation. It may be, however, that the propensity for late-onset diseases is associated with factors that do interfere with reproductive fitness in earlier life.

Several bacteria and viruses have been shown to possess antigens that resemble chemically the human A and B blood antigens. The smallpox virus, for example, has antigenic properties in common with group A blood cells. Accordingly, persons of blood group A would be expected to produce antibodies against the smallpox virus less effectively than persons lacking the A antigen. Prior to the advent of antibiotics and modern medical advances, infectious diseases resulted in high infant mortality. It could be, then, that the present-day frequencies of the blood-group alleles reflect the selective effects of past epidemics (such as, smallpox and plague). Even to this day, it is plausible that an individual's particular blood type might confer a selective advantage in combating certain bacteria and viruses.

Selected Readings

Blumberg, B. 1961. *Genetic polymorphisms and geographic variations in disease.* New York: Grune & Stratton.

Brues, A. M. 1954. Selection and polymorphism in the ABO blood groups. *American Journal of Physical Anthropology* 12:559–97.

Clarke, C. A. 1961. Blood groups and disease. In *Progress in medical genetics,* ed. A. G. Steinberg, vol. 1, pp. 81–119. New York: Grune & Stratton.

Clarke, C. A. 1968. The prevention of "rhesus" babies. *Scientific American,* November, pp. 46–52.

Crow, J. F., and Morton, N. E. 1960. The genetic load due to mother-child incompatibility. *American Naturalist* 94:413–19.

Levene, H., and Rosenfield, R. E. 1961. ABO incompatibility. In *Progress in medical genetics,* ed. A. G. Steinberg, vol. 1, pp. 120–57. New York: Grune & Stratton.

Levine, P. 1958. The influence of the ABO system on Rh hemolytic disease. *Human Biology* 30:14–28.

Newcombe, H. B. 1963. Risk of fetal death to mothers of different ABO and Rh blood types. *American Journal of Human Genetics* 15:499–564.

Race, R., and Sanger, R. 1968. *Blood groups in man.* Philadelphia: F. A. Davis Co.

Reed, T. E. 1971. Does reproductive compensation exist? An analysis of Rh data. *American Journal of Human Genetics* 23:215–24.

Stern, C. 1960. *Principles of human genetics.* San Francisco: W. H. Freeman and Co.

Vogel, F. 1970. ABO blood groups and disease. *American Journal of Human Genetics* 22:464–83.

9

Genetic Drift and Gene Flow

The peculiar multilegged condition of the bullfrog (see fig. 1.1) is a rarity in nature. Yet, we witnessed in a particular locality in 1958 an exceptionally high incidence of this defective trait. We have surmised that the deformity is caused by a recessive mutant gene. Since harmful recessive genes in a population tend to be carried mostly in the heterozygous state, the multilegged frogs probably arose from matings of heterozygous carriers. The probability that two or more carriers will actually meet is obviously greater in a small population than in a large breeding assemblage. In fact, the number of matings of carriers of a particular recessive gene in a population is mainly a function of the size of the population. Most populations are not infinitely large, and many fluctuate in size from time to time.

During a period when a population is small, chance matings and segregations could lead to an uncommonly high frequency of a given recessive gene. For example, it is not unthinkable that an unduly harsh winter sharply reduced the size of our particular bullfrog population. By sheer chance, an unusually large proportion of heterozygous carriers of the multilegged condition might have survived the winter's severity and prevailed as parents in the ensuing spring's breeding aggregation. In this manner, the "multilegged" gene, although not at all advantageous, would occur with an extraordinarily high incidence in the new generation of offspring. Such a fortuitous change in the genetic makeup of a population that may arise when the population becomes restricted in size is known as *genetic drift.*

Role of Genetic Drift

Examination of a natural situation demonstrating genetic drift will lead us into a simplified mathematical consideration of the concept. Coleman Goin, a naturalist at the University of Florida, studied the distribution of pigment variants of a terrestrial frog, known impressively as *Eleutherodactylus ricordi planirostris,* but commonly as the greenhouse frog. This frog may possess either of two pigmentary patterns, mottled or striped (fig. 9.1). A unique feature of the greenhouse frog is the terrestrial development of its eggs. That is, the eggs need not be submerged completely in water, but can develop in moist earth. This important quality may have considerable bearing on the dispersal of the frogs. Goin reared eggs successfully in a flowerpot two-thirds filled with beach sand and placed it in a finger bowl of water. An examination of a large number of progeny hatched from many different clutches of eggs revealed that the striped pattern is dominant to the mottled pattern.

The greenhouse frog is widespread in Cuba and the Bahama Islands, and has only recently become established in Florida. Cuba apparently has been the center of dispersal from which the Florida populations have been

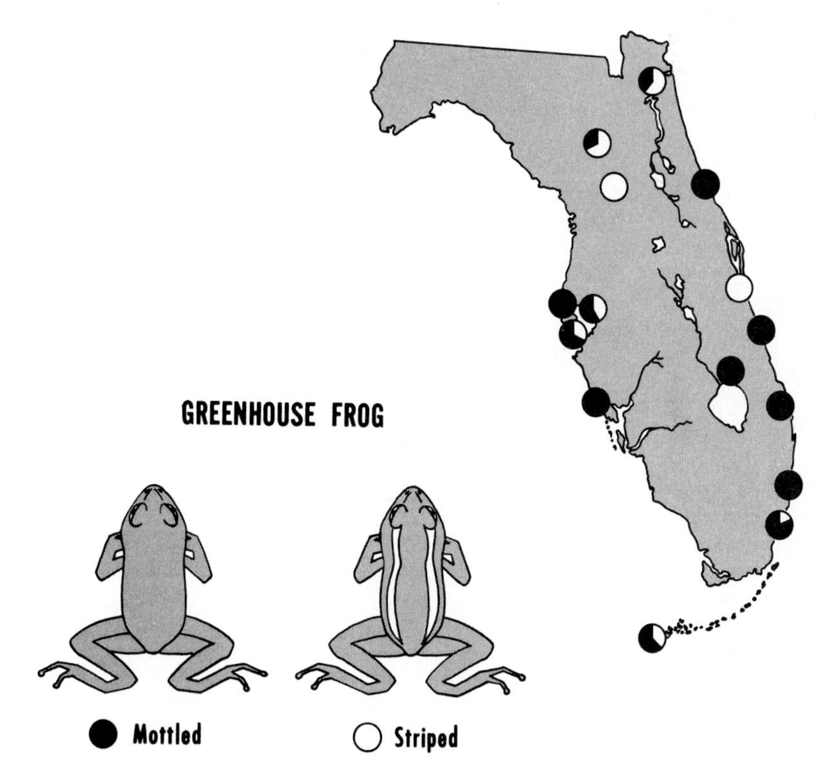

Figure 9.1 Distribution of the greenhouse frog in Florida, and the relative frequencies of the two pattern variants, *mottled* and *striped.* The populations are small and isolated, and differ appreciably in the incidences of mottled and striped forms. The varied frequencies may not be due to natural selection, but may represent the outcome of chance fluctuations of genes, or genetic drift. (Based on studies by Coleman Goin.)

derived. The present distribution of the greenhouse frog in Florida consists of a series of small, isolated colonies. As shown in figure 9.1, the proportions of the two patterns vary in different colonies in Florida. In several colonies, moreover, only the mottled type occurs. What can account for the local preponderance of one or the other pattern, or even the absence of one of the contrasting patterns?

Goin conjectures that the greenhouse frog was introduced into Florida by means of clutches of eggs accidentally included in shipments of plants from Cuba, a distinct possibility in view of the terrestrial development of the eggs. Thus, a single clutch of introduced eggs could initiate a small colony, which, in turn, would establish at the outset a given pattern or proportion of patterns. The presence of only mottled forms in a population may be due to the chance circumstance that only mottled eggs were introduced. Or, perhaps both striped and mottled eggs were included in the shipment, but by sheer accident, one type was lost in succeeding generations.

It should be understood that Goin has not proved that the unusual distribution and frequency of the two pigment patterns are due solely to chance. The demonstration of genetic drift in any natural population is an extremely difficult task. Genetic drift of the variant pigment patterns is, however, a reasonable explanation.

Theory of Genetic Drift

The theory of genetic drift was systematically developed in the 1930s by the geneticist Sewall Wright, then at the University of Chicago and later at the University of Wisconsin. In fact, the phenomenon of drift is frequently called the *Sewall Wright effect*. The Sewall Wright effect refers specifically to the random fluctuations (or drift) of gene frequencies from generation to generation in a population of small size. Because of the limited size of the breeding population, the gene pool of the new generation may not be at all representative of the parental gene pool from which it was drawn.

The essential features of the process of drift may be seen in the following modest mathematical treatment. Let us assume that the numerous isolated colonies in Florida were each settled by only two frogs, a male and a female, both having an *Aa* genotype. Let us further suppose that each mated pair produces only two offspring. The possible genotypes of the progeny, and the chance associations of the genotypes, are shown in table 9.1.

Several meaningful considerations emerge from table 9.1. For example, the chance that the first offspring from a cross of two heterozygous parents will be *AA* is ¼. The second event is independent of the first; hence, the chance that the second offspring will be *AA* is also ¼. The chance that *both* offspring will be *AA* is the product of the separate probabilities of the two independent events, ¼ x ¼, or 1/16.

We may now ask: What is the probability of producing two offspring,

Table 9.1

Chance Distribution of Offspring of Two Heterozygous Parents (Aa x Aa)

Genotype of first offspring	Probability of first event	Genotype of second offspring	Probability of second event	Total Probability
AA	1/4	*AA*	1/4	Both offspring *AA*, 1/16
AA	1/4	*Aa*	2/4	*AA* followed by *Aa*, 2/16
AA	1/4	*aa*	1/4	*AA* followed by *aa*, 1/16
Aa	2/4	*AA*	1/4	*Aa* followed by *AA*, 2/16
Aa	2/4	*Aa*	2/4	Both offspring *Aa*, 4/16
Aa	2/4	*aa*	1/4	*Aa* followed by *aa*, 2/16
aa	1/4	*AA*	1/4	*aa* followed by *AA*, 1/16
aa	1/4	*Aa*	2/4	*aa* followed by *Aa*, 2/16
aa	1/4	*aa*	1/4	Both offspring *aa*, 1/16

one *AA* and the other *Aa, in no particular order?* From table 9.1, we can see that the chance of obtaining an *AA* individual followed by an *Aa* individual is 2/16. Now, the wording of our question requires that we consider a second possibility, that of an *Aa* offspring followed by an *AA* offspring (also 2/16). These two probabilities must be added together to arrive at the chance of producing the two genotypes irrespective of the order of birth. Hence, in the case in question, the chance is 2/16 + 2/16, or 4/16. In like manner, it may be ascertained that the expectation of obtaining one *AA* and one *aa* offspring (in no given order) is 2/16; and that of producing one *Aa* and one *aa* (in any sequence) is 4/16.

The essential point is that any one of the above circumstances may occur in a given colony. We may concentrate on one situation. The probability that a colony will have only two *AA* offspring is 1 in 16. Thus, by the simple play of chance, the parents initiating the colony might not leave an *aa* offspring. The *a* gene would be immediately lost in the population. Subsequent generations descended from the first-generation *AA* individuals would contain, barring mutation, only *AA* types. Chance alone can thus lead to an irreversible situation. A gene once lost could not readily establish itself again in the population. The decisive factor is the size of the population. When populations are small, striking changes can occur from one generation to the next. Some genes may be lost or reduced in frequency by sheer chance; others may be accidentally increased in frequency. Thus, the genetic architecture of a small population may change irrespective of the selective advantage or disadvantage of a trait. Indeed, a beneficial gene may be lost in a small population before natural selection has had the opportunity to act on it favorably.

Founder Effect

When a few individuals or a small group migrate from a main population, only a limited portion of the parental gene pool is carried away. In the small migrant group, some genes may be absent or occur in such low frequency that they may be easily lost. The unique frequencies of genes that arise in populations derived from small bands of colonizers, or "founders," has been called the *founder effect.* This expression essentially emphasizes the conditions or circumstances that foster the operation of genetic drift.

The American Indians afford a possible example of the loss of genes by the founder principle. North American Indian tribes, for the most part, surprisingly have no I^B genes that govern type B blood. However, in Asia, the ancestral home of the American Indian, the I^B gene is widespread. The ancestral population of Mongoloids that migrated across the Bering Strait to North America might well have been very small. Accordingly, the possibility exists that none of the prehistoric immigrants happened to be of blood group B. It is also conceivable that a few individuals of the migrant band did carry the I^B gene but they failed to leave descendants.

The interpretation based on genetic drift should not be considered as definitive. The operation of natural selection cannot be flatly dismissed. Most of the North American Indians possess only blood group O, or stated another way, contain only the blood allele *i*. With few exceptions, the North American Indian tribes have lost not only blood group allele I^B but gene I^A as well. The loss of both alleles, I^A and I^B, by sheer chance perhaps defies credibility. Indeed, many modern students of evolution are convinced that some strong selective force led to the rapid elimination of the I^A and I^B genes in the American Indian populations. If this is true, it would offer an impressive example of the action of natural selection in modifying the frequencies of genes in a population.

North American Indians are also known to have a high frequency of albinism. The incidence of albinism among the Cuna Indians of the San Blas Province in Panama is about 1 in 200, which contrasts sharply with the 1 in 20,000 figure for European Caucasians. The Hopi Indians of Arizona and the Zuñi Indians of New Mexico, like the Cuna Indians, are also remarkable in their high numbers of albino individuals.

Since these American Indian populations are small, one might suspect the operation of genetic drift, notwithstanding the deleterious effects of the albino gene. It is difficult to imagine, however, that this detrimental gene could reach a high frequency on the basis of chance alone independently in several American Indian populations. Charles M. Woolf, the geneticist at

Arizona State University, has suggested that the high incidence of albinism among the Hopi Indians of Arizona reflects an inimitable form of selection that he terms *cultural selection.* Albinos have been highly regarded in the traditional Hopi society, and actually have enjoyed appreciable success in regard to sexual activity. The albino male has been admired, and some have become legendary for leaving large numbers of offspring. Woolf further notes that the fading of old customs among the Hopi Indians is beginning to nullify any reproductive advantage held by albino males in past generations. The frequency of albinism may be expected to decline with the dissolution of the traditional Hopi way of life.

Religious Isolates

The most likely situation to witness the phenomenon of genetic drift is one in which the population is virtually a small, self-contained breeding unit, or *isolate,* in the midst of a larger population. This typifies the Dunkers, a very small religious sect in eastern Pennsylvania. The Dunkers are descendants of the Old German Baptist Brethren, who came to the United States in the early 18th century. Bentley Glass, then at Johns Hopkins University, studied the community of Dunkers in Franklin County, Pennsylvania, which numbers about 300 individuals. In each generation, the number of parents has remained stable at about 90. The Dunkers live on farms intermingled with the general population, but are genetically isolated by rigid marriage customs. The choice of mates is restricted to members within the religious group.

Glass, with his colleagues, compared the frequencies of certain traits from the Dunker community, the surrounding heterogeneous American population, and the population in western Germany from which the Dunker sect had emigrated two centuries ago. Such a comparison of a small isolate with its large host and parent populations should reveal the effectiveness, if any, of genetic drift. In other words, if the small isolated population shows aberrant gene frequencies as compared to the large parent population, and if the other forces of evolution can be excluded, then the genetic differences can be ascribed to drift.

Analyses were made of the patterns of inheritance of three blood group systems—the ABO blood groups, the MN blood types, and the Rh blood types. In addition, data were accumulated on the incidences of four external traits—namely, the configuration of the ear lobes (which either may be attached to the side of the head or hang free), right- or left-handedness, the presence or absence of hair on the middle segments of the fingers (mid-digital hair), and "hitch-hiker's thumb," technically termed distal hyperextensibility (fig. 9.2).

The frequencies of many of these traits are strikingly different in the Dunker community from those of the general United States and West Ger-

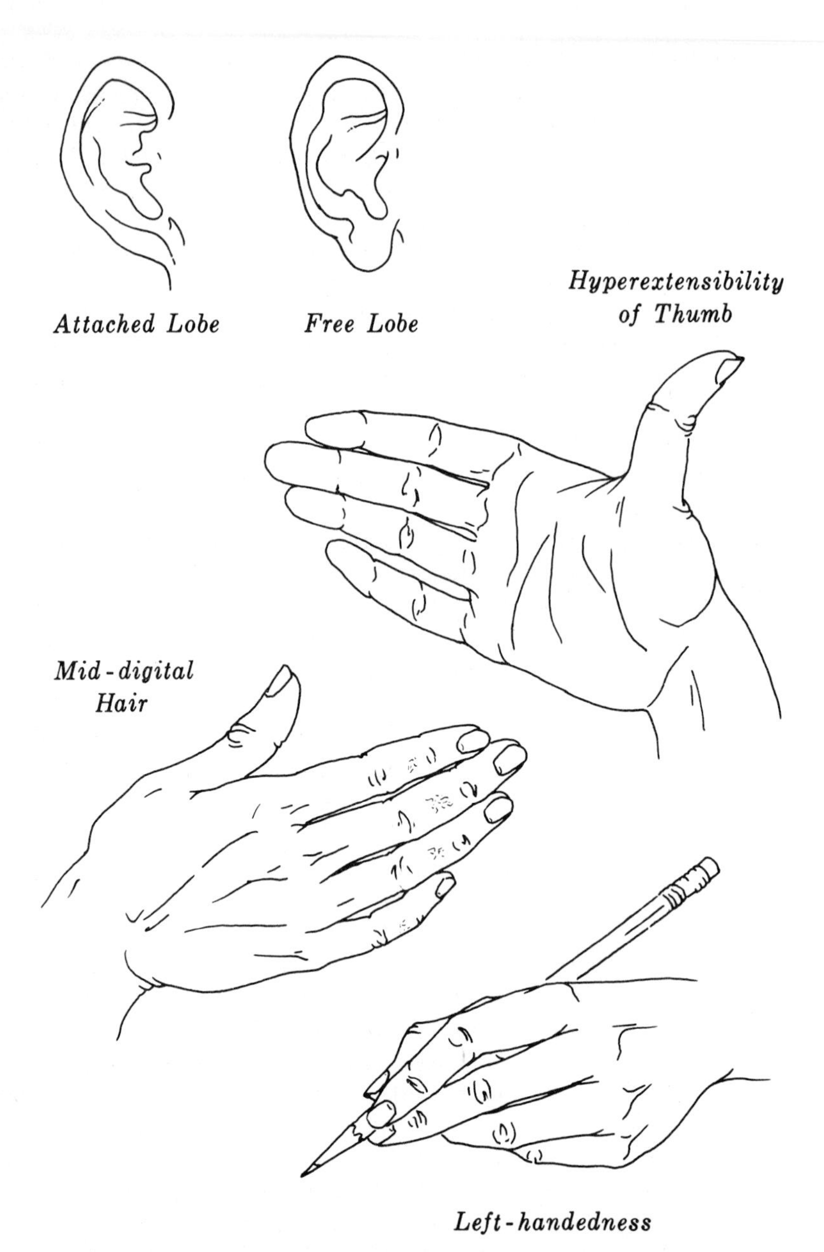

Figure 9.2 Inheritable physical traits—nature of ear lobes, "hitch-hiker's thumb," mid-digital hair, and handedness—studied by Bentley Glass and his co-workers in members of the small religious community of Dunkers in Pennsylvania. The distinctive frequencies of most of these traits in the Dunker population suggests the operation of genetic drift.

man populations. Blood group A is much more frequent among the Dunkers; the O group is somewhat rarer in the Dunkers; and the frequencies of groups B and AB have dropped to exceptionally low levels in the Dunker community. In fact, the I^B gene had almost been lost in the isolate. Most of the carriers of the I^B gene were not born in the community, but were converts who entered the isolate by marriage.

A noticeable change has occurred also in the incidences of the M and N blood types in the Dunker community. Type M has increased in frequency, and type N has dwindled in frequency as compared with the incidences of these blood types in either the general United States population or the West German population. Only in the Rh blood groups do the Dunkers conform closely to their surrounding large population.

In physical traits, equally striking differences were found. Briefly, the frequencies of mid-digital hair patterns, distal hyperextensibility of the thumb, and attached ear lobes are significantly lower in the Dunker isolate than in the surrounding American populations. The Dunkers do, however, agree well with other large populations in the incidence of left-handedness. It would thus appear that the peculiar constellation of gene frequencies in the Dunker community—some uncommonly high, others uniquely low, and still others, unchanged from the general large population—can be best attributed to chance fluctuations, or genetic drift.

There is no concurrence of opinion among evolutionists concerning the operation of genetic drift in natural populations, but few would deny that small religious isolates have felt the effect of random sampling. It should be clear, however, that genetic drift becomes ineffectual when a small community increases in size. Fluctuations or shifts in gene frequencies in large populations are determined almost exclusively by selection.

Amish of Pennsylvania

We have seen that gene frequencies in small religious isolates may differ significantly from the original large populations from which the isolates were derived. Another feature of small isolates is the occurrence of rare recessive traits in greater numbers than would be expected from random mating in a large population. This is witnessed among the Old Order Amish societies in the eastern United States.

The Amish sect is an offshoot of the Mennonite Church; both religious groups settled in the United States to escape persecution in Europe in past centuries. The Amish are old-fashioned, rural-living people who cultivate the religious life apart from the world. Present-day communities were founded by waves of Amish immigration that began about 1720 and continued until about 1850. The vast majority of Amish live in relatively isolated colonies in Pennsylvania, Ohio, and Indiana. Each community is descended from a small immigrant stock, as attested by the relatively few family names in a given community. Analyses by the geneticist Victor A.

McKusick of Johns Hopkins University have shown that eight names account for 80 percent of the Amish families in Lancaster County, Pennsylvania. Other Amish communities also are characterized by a high frequency of certain family names, as table 9.2 shows.

Table 9.2

Old Order Amish Family Names in Three American Communities*

Lancaster Co., Pa.		Holmes Co., Ohio		Mifflin Co., Pa.	
Stolzfus**	23%	Miller	26%	Yoder	28%
King	12%	Yoder	17%	Peachey	19%
Fischer	12%	Troyer	11%	Hostetler	13%
Beiler	12%	Hershberger	5%	Byler	6%
Lapp	7%	Raber	5%	Zook	6%
Zook	6%	Schlabach	5%	Speicher	5%
Esh***	6%	Weaver	4%	Kanagy	4%
Glick	3%	Mast	4%	Swarey	4%
	81%		77%		85%
Totals:					
1,106 families, 1957		1,611 families, 1960		238 families, 1951	

* From data compiled by Victor A. McKusick of Johns Hopkins University.
** Including Stolzfoos.
*** Including Esch.

Marriages have been largely confined within members of the Amish sect, with a resulting high degree of consanguinity. Marriages of close relatives have tended to promote the meeting of two normal, but carrier, parents. Four recessive disorders manifest themselves with uncommonly high frequencies, each in a different Amish group: the Ellis-van Creveld syndrome, pyruvate-kinase-deficient hemolytic anemia, Hemophilia B (Christmas disease), and a form of limb-girdle muscular dystrophy (Troyer syndrome).

We may consider in some detail the Ellis-van Creveld syndrome, which occurs in the Lancaster County population (see fig. 1.5). Fifty-two affected persons have been identified in 30 sibships, most of which have unaffected parents. Pedigree analysis has revealed that Samuel King and his wife, who immigrated in 1744, are ancestral to all parents of the sibships. Either Samuel King or his wife carried the recessive gene. None of their children were affected, but subsequent generations were. Evidently, previously concealed detrimental recessive genes are brought to light by the increased chances of two heterozygotes meeting in a small population.

Protein Evolution and Genetic Drift

We had earlier learned (chapter 3) that a gene (a linear array of bases in the DNA) codes for the precise sequence of amino acids that compose a protein. The substitution of one amino acid for another is traceable typically to a single point mutation—that is, an alteration of a single base of the triplet of DNA that specifies the amino acid. Single amino acid substitutions that interfere with the function of the protein, or decrease drastically the rate of synthesis of the protein, are likely to be discarded by natural selection.

We now have substantial knowledge of the amino acid sequence of cytochrome *c*, a respiratory enzyme containing about 100 amino acids. This protein has been analyzed in more than 30 species of organisms, and the same amino acids have been found at 20 positions in all organisms tested from mold to man. Apparently, the 20 amino acids at specific positions are irreplaceable, and any substitutions at these sites are likely to interfere with the function of the enzyme. In the remaining portion of the chain, however, several different amino acids appear at a given position. The variable nature of these remaining amino acids could be interpreted as indicating that amino acid replacements at some positions are irrelevant for the function of the protein. That is, amino acid substitutions at certain positions are not likely to provide either an evolutionary advantage or disadvantage, and can be preserved by chance. In essence, it may be that certain amino acid replacements result from neutral mutations that are fixed by genetic drift.

The notion that amino acid substitutions may be fixed by the random drift of neutral mutations remains controversial. Advocates point to the analysis of fibrinopeptide A, a polypeptide concerned with the blood-clotting mechanism. Among several mammalian species, there is an extraordinarily high level of amino acid substitutions in fibrinopeptide A. This polypeptide presumably functions equally well with numerous different amino acid sequences. It has been argued that the mutations responsible for the many different amino acids must be neutral. In other cases, however, it is difficult to deny the role of natural selection. Certain histones (which form complexes with DNA) are highly conserved in their amino acid sequences. For example, histone IV from such divergent organisms as the pea plant and the calf differ in only two of 102 amino acid residues. All other amino acid substitutions apparently would disrupt the activity of histone IV, and such substitutions that may have appeared in the evolutionary past were probably eliminated by natural selection.

Gene Flow

A rich archeological record reveals appreciable movement on the part of early human populations. Some migrations were sporadic, in small groups;

others were more or less continual streams, involving large numbers of peoples. Large-scale immigrations followed by interbreeding have the effect of introducing new genes to the host populations. The diffusion of genes into populations through migrations is referred to as *gene flow*.

The graded distribution of the I^B blood-group gene in Europe represents the historical consequence of invasions by Mongolians who pushed westward repeatedly between the 6th and 16th centuries (fig. 9.3). There is a high frequency of the I^B gene in central Asia. In Europe, the frequency of the I^B gene diminishes steadily from the borders of Asia to a low level of 5 percent or less in parts of Holland, France, Spain, and Portugal. The Basque peoples, who inhabit the region of the Pyrenees in Spain and France, have the lowest frequency of the I^B gene in Europe—below 3 percent. From a biological standpoint, the Basque community of long standing is a cohesive, endogamous mating unit. The exceptionally low incidence of the I^B gene among the Basques may be taken to indicate that there has been little intermarriage with surrounding populations. It is possible that a few centuries ago the I^B gene was completely absent from the self-contained Basque community.

The Basques are also of interest in having a high frequency of the Rh-negative allele, *r* (see chapter 8). These peoples may be the descendants of an early European population that was almost exclusively *rr*. The Rh-

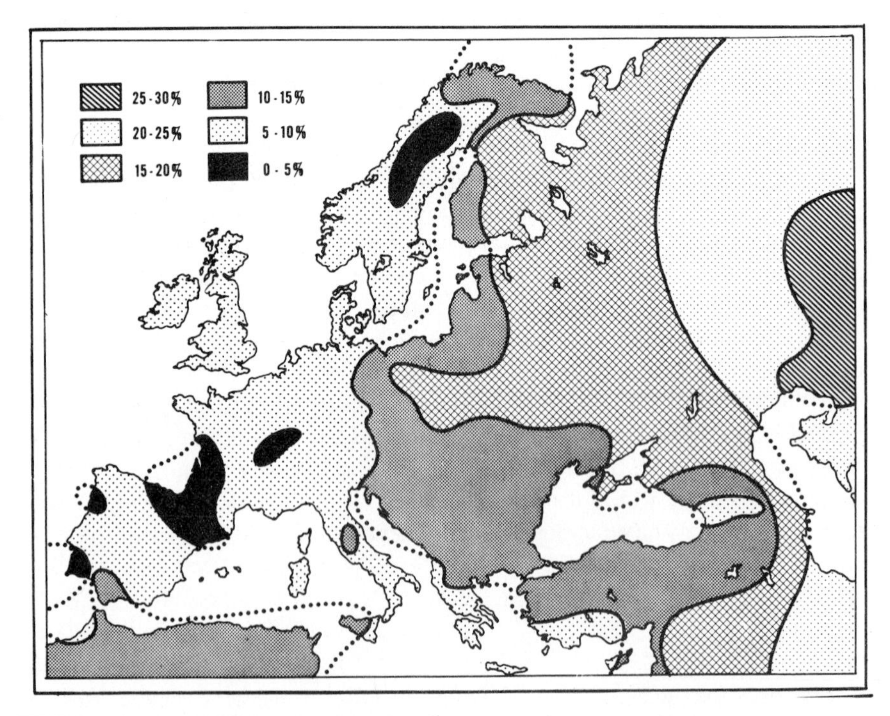

Figure 9.3 Gradient of frequencies of the I^B blood-group gene from central Asia to western Europe. (Based on studies by A. E. Mourant.)

positive allele (*R*) was probably introduced by migrants into the Basque community, much in the same way that the Rh-negative allele (*r*) is presently being introduced in populations of China. Until recently, Rh disease was virtually unknown in China. All Chinese women are Rh-positive (*RR*). However, intermarriage between immigrant Americans and the native Chinese has led to the introduction of the Rh-negative gene (*r*) in the Chinese population. No Rh disease would be witnessed in the immediate offspring of American men and Chinese women. Conversely, all marriages of Rh-negative American women (*rr*) and Rh-positive men (*RR*) would be of the incompatible type. All children by Chinese fathers would be Rh-positive (*Rr*) and potential victims of hemolytic disease.

Whereas American immigrants introduce the Rh-negative gene (*r*) into Chinese populations where it formerly was not present, Chinese immigrants (all of whom are *RR*) introduce more Rh-positive genes (*R*) into the American populations, thus diluting the Rh-negative gene pool in the United States. Initially, the Rh-positive Chinese men (*RR*) married to Rh-negative American women (*rr*) would result in an increased incidence of Rh-diseased infants. In later generations, however, the frequency of Rh-negative women in the United States would be lower, inasmuch as women of mixed Chinese-American origin would be either *RR* or *Rr*, predominantly the former. Thus, in the United States, the long-range effect of Chinese-American intermarriage is a reduction in the incidence of hemolytic disease of the newborn.

Caucasian Genes in American Negroes

In the United States, from a beginning more than 250 years ago, there has occurred an admixture of Caucasians with Negroes who were brought from Africa to the New World as slaves. The overall Caucasian contribution to the American Negro gene pool has been estimated at approximately 25 percent. This percentage must be considered as a gross estimate, inasmuch as different Negro populations in the United States have undergone varied degrees of admixture.

The geneticist T. Edward Reed of the University of Minnesota has measured the frequency with which a particular blood-group gene appears in Caucasians, African Negroes, and American Negroes. The specific gene is the Fy^a gene (the Duffy factor) of the Duffy blood group system. As seen in table 9.3, the Duffy factor is virtually absent in the African Negro populations from which most of the original immigrant slaves were derived. The highest frequency of the Duffy factor in the stem populations in Africa is an exceedingly low 0.04, whereas the value for representative Caucasian populations in the United States is 0.4, or ten times as great (table 9.3). Among Negroes residing in New York City, Detroit, and Oakland, the gene reaches a frequency of about 0.10. This signifies that the magnitude of Caucasian ancestry in Negroes in these localities may be as high as 26 percent. In con-

Table 9.3

Duffy Blood Factor (Fy^a Gene) in Caucasian, African Negro, and American Negro Populations*

Region	Frequency of Duffy Factor	Percentage of Caucasian Contribution
A. African Negro		
Liberia	0.00	—
Ivory Coast	0.04	—
Upper Volta	0.00	—
Dahomey	0.00	—
Ghana (Accra)	0.00	—
Nigeria (Lagos)	0.00	—
B. Caucasian		
Oakland, California	0.43	
Evans and Bullock Counties, Georgia	0.42	
C. American Negro		
1. Non-Southern		
New York, New York	0.08	18.9
Detroit, Michigan	0.11	26.0
Oakland, California	0.09	22.0
2. Southern		
Charleston, South Carolina	0.02	3.7
Evans and Bullock Counties, Georgia	0.05	10.6

* Based on data compiled by T. Edward Reed of the University of Minnesota.

trast, the frequency of the Duffy factor among Negroes in Charleston, South Carolina is 0.02, which indicates that the Negro population in Charleston has only a small amount of Caucasian ancestry. Thus, although the extent of Caucasian-Negro hybridization has been variable, it is clear that various American Negro populations derive between 4 percent and 26 percent of their genes from Caucasian ancestors, and these Caucasian genes have been introduced through hybridization since 1700.

Selected Readings

Boyd, W. C. 1963. Four achievements of the genetical method in physical anthropology. *American Anthropologist* 65:243–52.

Candela, P. B. 1942. The introduction of blood-group B into Europe. *Human Biology* 14:413–43.

Cavalli-Sforza, L. L. 1969. "Genetic drift" in an Italian population. *Scientific American,* August, pp. 30–37.

Cavalli-Sforza, L. L. Barrai, I., and Edwards, A. W. F. 1964. Analysis of human evolution under random genetic drift. *Cold Spring Harbor Symposium on Quantitative Biology* 29:9–20.

Dunn, L. C. 1966. *Heredity and evolution in human populations.* New York: Atheneum Publishers.

Dunn, L. C., and Dunn, S. P. 1957. The Jewish community of Rome. *Scientific American,* March, pp. 118–28.

Glass, B. 1953. The genetics of the Dunkers. *Scientific American,* August, pp. 76–81.

———. 1954. Genetic changes in human populations, especially those due to gene flow and genetic drift. *Advances in Genetics* 6:95–139.

———. 1956. On the evidence of random genetic drift in human populations. *American Journal of Physical Anthropology* 14:541–55.

Glass, B., and Li, C. C. 1953. The dynamics of racial intermixture—an analysis based on the American Negro. *American Journal of Human Genetics* 5:1–20.

Goldsby, R. A. 1971. *Race and races.* New York: Macmillan Co.

Goldschmidt, E. 1963. *The genetics of migrant and isolate populations.* Baltimore: Williams & Wilkins Co.

McKusick, V. A., Hostetler, J. A., Egeland, J. A., and Eldridge, R. 1964. The distribution of certain genes in the Old Order Amish. *Cold Spring Harbor Symposium on Quantitative Biology* 29:99–114.

Mourant, A. E. 1954. *The distribution of the human blood groups.* Oxford, England: Blackwell Scientific Publications.

Pollitzer, W. S. 1958. The Negroes of Charleston (S.C.): a study of hemoglobin types, serology and morphology. *American Journal of Physical Anthropology* 16:241–63.

Reed, T. E. 1969. Caucasian genes in American Negroes. *Science* 165:762–68.

Woolf, C. M., and Dukepoo, F. C. 1969. Hopi Indians, inbreeding, and albinism. *Science* 164:30–37.

Wright, S. 1951. Fisher and Ford on "The Sewall Wright Effect." *American Scientist* 39:452–58.

10

Races and Species

Any large assemblage of a particular organism is generally not distributed equally nor uniformly throughout its territory or range in nature. A widespread group of plants or animals is typically subdivided into numerous local populations, each physically separated from the others to some extent. The environmental conditions in different parts of the range of an organism are not likely to be identical. We may thus expect that a given local population will consist of genetic types adapted to a specific set of prevailing environmental conditions. The degree to which each population maintains its genetic distinctness is governed by the extent to which *interbreeding* between the populations occurs. A free interchange of genes between populations tends to blur the differences between the populations. But what are the consequences when gene exchange between populations is greatly restricted or prevented? This chapter addresses itself to this question.

Variation Between Populations

Our first consideration is to demonstrate that inheritable variations exist among the various breeding populations in different geographical localities of an organism. Jens Clausen, David Keck, and William Hiesey of the Carnegie Institution of Washington at Stanford, California, have shown that each of the populations of the yarrow plant, *Achillea lanulosa,* from different parts of California is adapted to its respective habitat. As figure 10.1 shows, the variations in height of the plant are correlated with altitudinal differences. The shortest plants are from the highest altitudes, and the plants increase in height in a gradient fashion with decreasing altitude. The term *cline,* or character gradient, has been applied to such situations where a character varies more or less continuously with a gradual change in the environmental terrain.

The observation by itself that the yarrow plants are phenotypically dissimilar at different elevations does not indicate that they are genetically different. If the observed variations are claimed as local adaptations resulting from natural selection, then a hereditary basis for the differences in height should be demonstrated. It is often difficult to obtain data that disclose the hereditary nature of population differences. In this respect, the studies of Clausen and his co-workers are commendable. The plants shown in figure 10.1 had actually been grown together in a uniform experimental garden at Stanford, California. The plants, transplanted from various localities, developed differently from one another in the same experimental garden, revealing that each population had evolved its own distinctive complex of genes.

Races

The variation pattern in organisms may be discrete, or discontinuous, particularly when the populations are separated from each other by pronounced

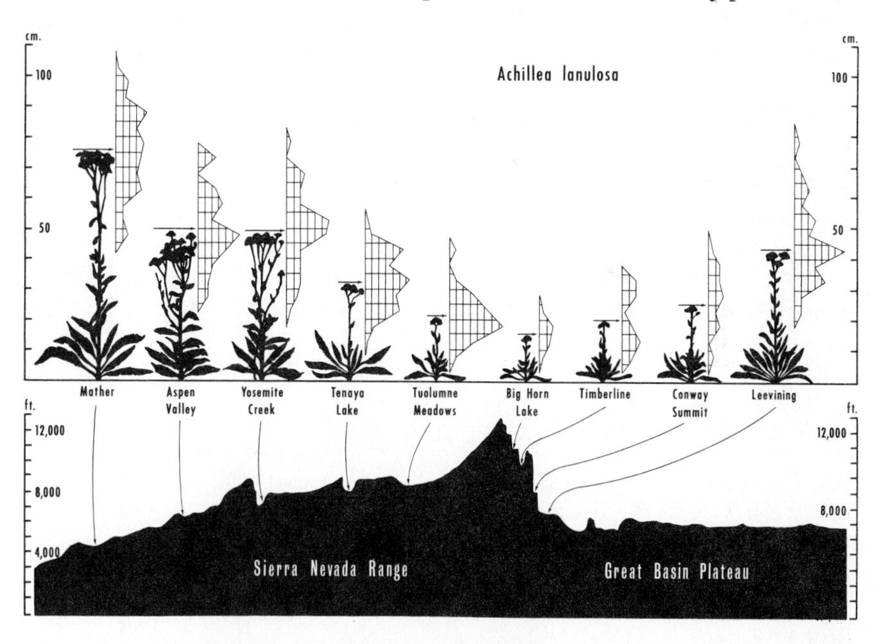

Figure 10.1 Clinal variation in the yarrow plant, *Achillea lanulosa.* The increase in height of the plant is more or less continuous with decreasing altitude. The plants shown here are representatives from different populations in the Sierra Nevada Mountains of California that were grown in a uniform garden at Stanford, California. Each plant illustrated is one of average height for the given population; the graph adjacent to the plant reveals the distribution of heights within the population. (From Clausen, Keck, and Hiesey, *Carnegie Institution of Washington Publication 581,* 1948.)

physical barriers. This is exemplified by the varieties of the carpenter bee (*Xylocopa nobilis*) in the Celebes and neighboring islands of Indonesia (fig. 10.2). As shown by the studies of J. van der Vecht of the Museum of Natural History at Leiden in the Netherlands, there are three different varieties on the mainland of Celebes, and at least three kinds on the adjacent small islands. These geographical variants differ conspicuously in the coloration of the small, soft hairs that cover the surface of the body. The first abdominal segment is invariably clothed with bright yellow hairs. However, each variety has evolved a unique constellation of colors on the other abdominal segments and also on the thorax.

The variations in the carpenter bees within and between islands are

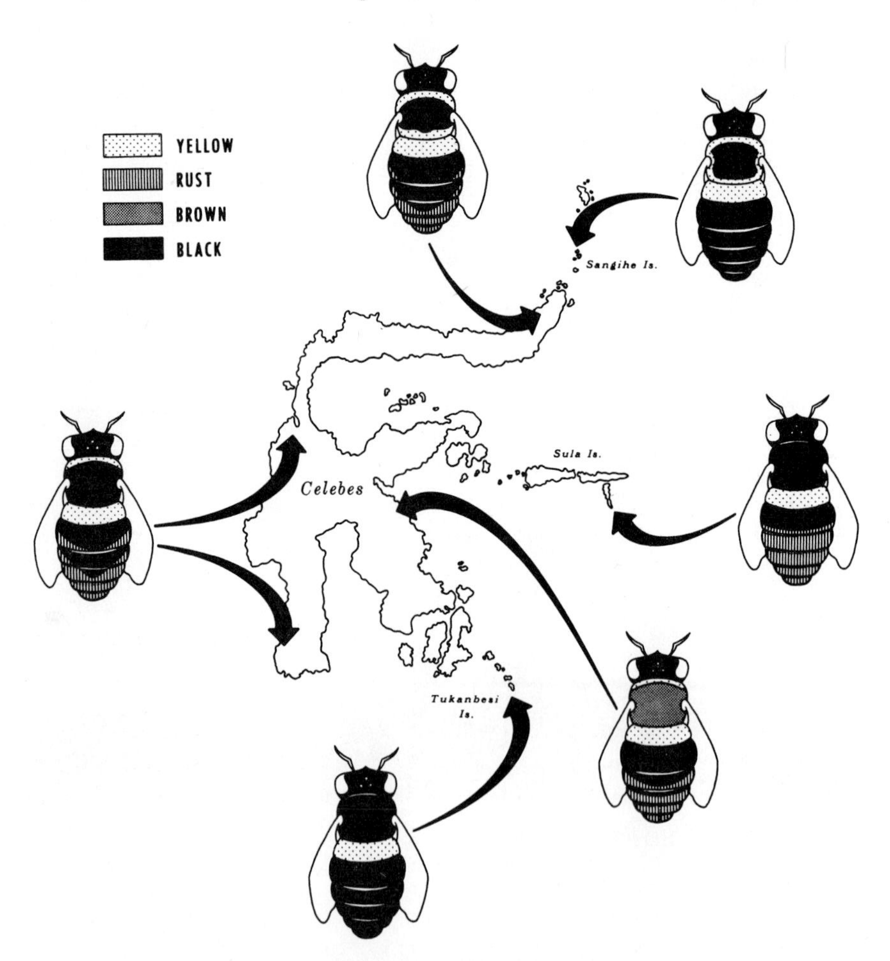

Figure 10.2 Geographic variation of color patterns in females of the carpenter bee, *Mylocopa nobilis,* in the Celebes and neighboring islands in Indonesia. Each geographic race has evolved a distinctive constellation of colors. (Based on studies by J. van der Vecht.)

well defined and easily distinguishable. One may refer to populations with well-marked discontinuities as *races*. Races are simply geographical aggregates of populations that differ in the incidence of genetic traits. How genetically different two assemblages of populations must be to warrant racial designations is an open question.

Some of the problems inherent in delimiting races are exemplified by the different temperature-adapted populations of the North American leopard frog, *Rana pipiens*. John A. Moore, then at Columbia University and later at the University of California, tested the effects of temperature on the development of the embryos of frogs from widely different localities. He wished to ascertain the limits of temperatures that the embryos can endure or tolerate. The findings on four different geographic populations in the eastern United States are shown in figure 10.3.

Embryos of northern *Rana pipiens* populations are more resistant to low temperatures and less tolerant of high temperatures than are embryos from southern populations. Embryos of populations from Vermont and New Jersey have comparable ranges of temperature tolerances. These northern embryos can resist temperature as low as 5°C. Embryos from Florida differ markedly from those of northern populations. Embryos from southern Florida (latitude 27°N) can tolerate temperatures as high as 35°C, but are very susceptible to low temperatures. Hence, northern and southern populations have become adapted to different environments in their respective territories.

We may refer to the northern populations as the cold-adapted race of the leopard frog, and designate the southern populations as the warm-adapted race. It is evident that we are being arbitrary in drawing a fine line of demarcation between northern and southern races. Data are presently lacking for the geographically intermediate populations, but further studies will probably reveal that the temperature adaptations of the frog embryos change gradually from north to south. Even with the present information, one may wish to recognize more than two temperature-adapted races, or perhaps, as some investigators firmly argue, refrain completely from making racial designations.

Races may be best thought of as units of organization below the species level. In other words, races may be considered as stages in the transformation of populations into species. But what constitutes a species? Up to this point we have assiduously avoided the use of the term *species*. A discussion of the process leading to the formation of species will facilitate understanding of the term itself.

Formation of Species

Let us imagine a large assemblage of land snails subdivided in three geographical aggregations or races, *A*, *B*, and *C*, each adapted to local environ-

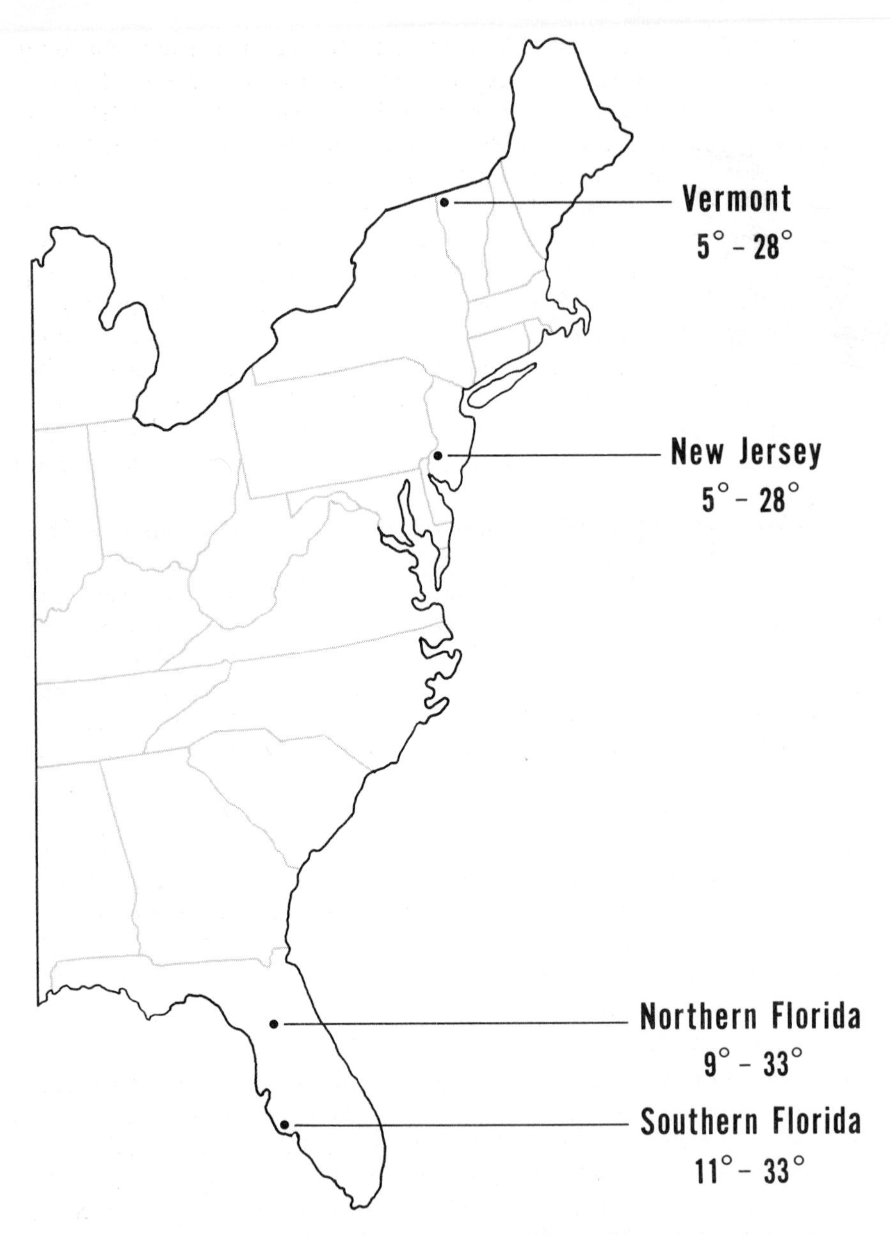

Figure 10.3 Limits of temperature tolerance of embryos of the leopard frog (*Rana pipiens*) from different geographical populations. Embryos of northern populations are more resistant to low temperatures and less tolerant of high temperatures than are embryos from southern populations. (Based on studies by John A. Moore.)

mental conditions (fig. 10.4). There are initially no gross barriers separating the populations from each other, and where *A* meets *B* and *B* meets *C*, interbreeding occurs. Zones of intermediate individuals are thus established between the races, and the width of these zones depends on the extent to which the respective populations intermingle. It is important to realize that races are fully capable of exchanging genes with one another.

We may now visualize (fig. 10.4) some striking physical feature, such as a great river, forging its way through the territory and effectively isolating the land snails of race *C* from those of *B*. These two assemblages may be spatially separated from each other for an indefinitely long period of time, affording an opportunity for race *C* to pursue its own independent evolutionary course. Two populations that are geographically separated, like *B* and *C* in our pictorial model, are said to be *allopatric*. (Technically speaking, *A* and *B* are also allopatric, since they, for the most part, occupy different geographical areas.)

After eons of time, the river may dry up and the hollow bed may eventually become filled in with land. Now, if the members of populations *B* and *C* were to extend their ranges and meet again, one of two things might happen. The snails of the two populations may freely interbreed and establish once again a zone of intermediate individuals. On the other hand, the two populations may no longer be able to interchange genes. If the two assemblages can exist side by side without interbreeding, then the two groups have reached the evolutionary status of separate species. *A species is a breeding community that preserves its genetic identity by its inability to exchange genes with other such breeding communities.* In our pictorial model (fig. 10.4), race *C* has became transformed into a new species, *C′*. Two species (*A-B* and *C′*) have now arisen where formerly only one existed. It should be noted that races *A* and *B* are treated as members of a single species since no barriers to gene exchange exist between them.

Nomenclature

The scientific names that the taxonomist would apply to our populations of land snails deserve special comment. The technical name of a species consists of two words, in Latin or in latinized form. An acceptable designation of the original species of land snails depicted in figure 10.4 would be *Helix typicus*. The first word is the name of a comprehensive group, the genus, to which land snails belong; the second word is a name unique to the species. The taxonomist would be obliged to create a different latinized second name for the newly derived species of land snail, the *C′* population in figure 10.4. This new species might well be called *Helix varians*. The name of the genus remains the same since the two species are closely related. The genus, therefore, denotes a group of interrelated species.

The binomial ("two-named") system of nomenclature, universally

Geographical Variation

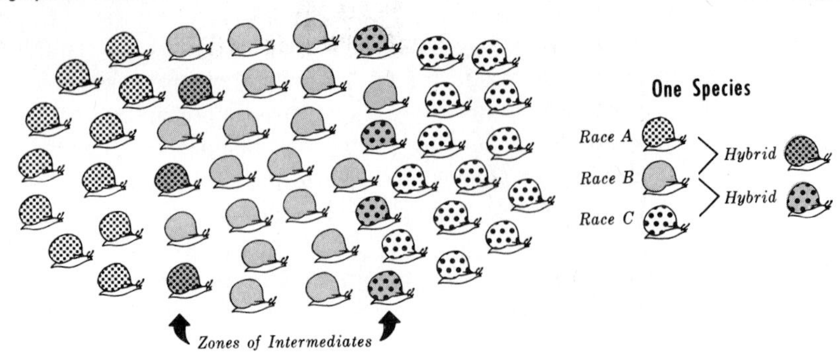

Isolating Action of Geographic Barrier

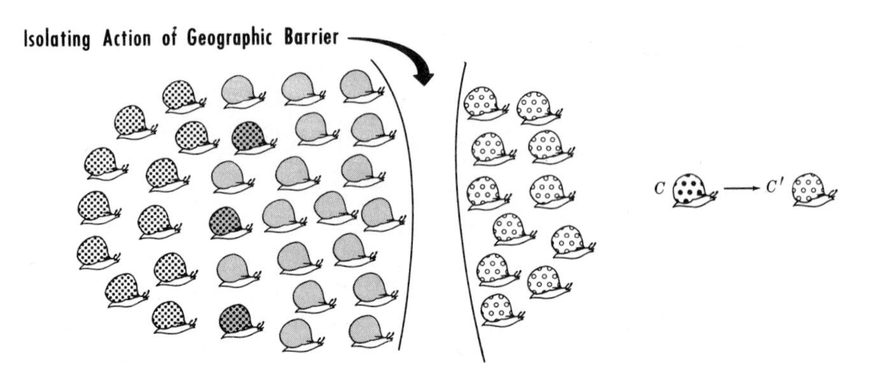

Removal of Geographic Barrier

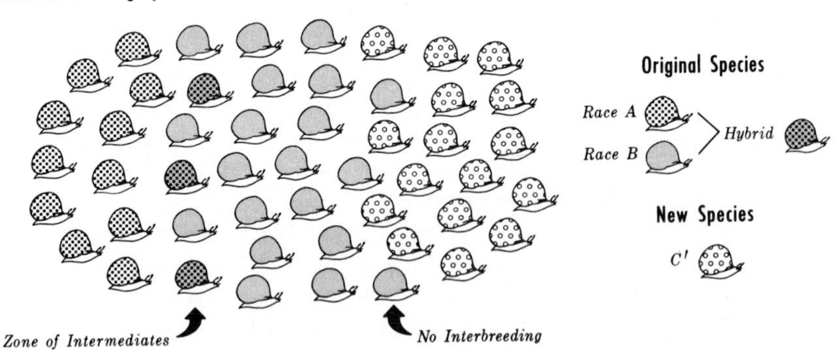

Figure 10.4 Model for the process of geographic speciation. Members of population (or race) C had diverged genetically during geographical isolation in ways that have made them reproductively incompatible with race B when they met again. Race C has thus transformed into a new species, C′.

accepted, was devised by the Swedish naturalist Carolus Linnaeus (born Karl von Linné) in his monumental work, *Systema Naturae,* first published in 1735. Convention dictates that the first letter of the generic name be capitalized and that the specific name begin with a small letter. It is also customary to print the scientific name of a species in italics, or in a type that is different from that of the accompanying text. A modern refinement of the Linnaean system is the introduction of a third italicized name, which signifies the subspecies. Geographical races are recognized taxonomically as subspecies. Thus, it would be appropriate to designate races *A* and *B* (fig. 10.4) as *Helix typicus elegans* and *Helix typicus eminens,* respectively. Such a species composed of two (or more) subspecies is said to be *polytypic.* A monotypic species is one that is not differentiated into two or more geographical races or subspecies. *Helix varians* would be a monotypic species.

Reproductive Isolating Mechanisms

We have seen that two populations (or races), while spatially separated from each other, may accumulate sufficient genetic differences in isolation to prevent an interchange of genes if they came into contact with one another. When the geographical barrier persists, it is difficult to judge the extent of genetic divergence between the two allopatric populations. Only when the two populations come together again does it become apparent whether or not they have changed in ways that would make them reproductively incompatible. Two populations that come to occupy the same territory are called *sympatric.* The ways or agencies that prevent interbreeding between sympatric species are known as *reproductive isolating mechanisms.*

Reproductive isolating mechanisms take varied forms, and one or more of the different types may be found separating two species. The various types may be grouped into two broad categories. One category includes the prezygotic (or premating) mechanisms, which serve to prevent the formation of hybrid zygotes. The other category encompasses the postzygotic (or postmating) mechanisms, which act to reduce the viability or fertility of hybrid zygotes. The specific types of isolating mechanisms under these two groupings can be listed as follows:

A. Prezygotic (premating) mechanisms
   1. Habitat (ecological) isolation
   2. Seasonal (temporal) isolation
   3. Sexual (ethological) isolation
   4. Mechanical isolation
   5. Gametic isolation

B. Postzygotic (postmating) mechanisms
   1. Hybrid inviability
   2. Hybrid sterility
   3. Hybrid breakdown

Two related species may live in the same general area, but differ in their ecological requirements. The scarlet oak (*Quercus coccinea*) of eastern North America grows in moist or swampy soils, whereas the black oak (*Quercus velutina*) is adapted to drier soils (fig. 10.5). The two kinds of

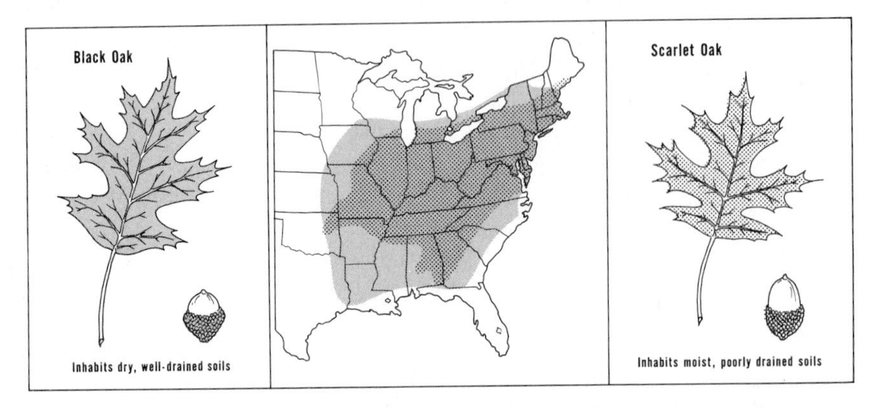

Figure 10.5 Ecological isolation is exemplified by the different habitat requirements of two species of oaks. Although both species occur together in eastern United States, the scarlet oak (*Quercus coccinea*) is adapted to the moist bottom lands, whereas the black oak (*Quercus velutina*) is adapted to the dry upland soils.

oaks are thus effectively separated by different *ecological* or *habitat* preferences. Two sympatric species may also retain their distinctness by breeding at different times of the year (*seasonal isolation*). Evidently, cross-fertilization is not feasible between two species of frogs that release their gametes on different months even in the same pond, or between two species of pine that shed their pollen in different periods. The breeding seasons of two species may overlap, but interbreeding may not occur because of the lack of mutual attraction between the sexes of the two species (*sexual isolation*). Among birds, for example, elaborate courtship rituals play important roles in species recognition and the avoidance of interspecific matings.

In many insects, interbreeding of species is hindered by differences in the structure of the reproductive apparatus (*mechanical isolation*). Copulation is not possible because the genitalia of one species is physically incompatible with the genitalia of the other species. In fact, several closely related species of insects can often be accurately classified by their distinctive genitalia. In some instances, the male of one species may inseminate the female of the other species, but the sperm cells may be inviable in the reproductive tract of the female. This form of *gametic isolation* is not unique to animals; in plants, such as the Jimson weed (*Datura*), the sperm-bearing pollen tube of one species encounters a hostile environment in the flower tissue of the other species and is unable to reach the egg.

Cross-fertilizations between two species may be successful, but the hybrid embryos may be abnormal or fail to reach sexual maturity (*hybrid inviability*). For example, two species of the chicory plant, *Crepis tectorum* and *Crepis capillaris,* can be crossed, but the hybrid seedlings die in early development. Crosses between the bullfrog, *Rana catesbeiana,* and the green frog, *Rana clamitans,* result in inviable embryos. In certain hybrid crosses, such as between females of the toad species *Bufo fowleri* and males of *Bufo valliceps,* the hybrids may survive but are completely sterile (*hybrid sterility*). The familiar example of hybrid sterility is the mule, the offspring of a male ass and a female horse. In some situations, the F_1 hybrids appear to be vigorous and fertile, but the viability of a subsequent generation is very reduced. Such a case of *hybrid breakdown* has been described in several species of cotton—*Gossypium hirsutum* and *Gossypium barbadense.* They produce normal, fertile F_1 hybrids, but the majority of the F_2 hybrid cotton seedlings fail to germinate.

In essence, two populations can remain genetically distinct, and be designated as species, then gene exchange between them is prevented or limited by one or more reproductive isolating mechanisms. More often than not, we are unable to obtain direct evidence for the presence or absence of interbreeding in nature between two groups. The degree of reproductive isolation is then indirectly gauged by the extent to which the members of two populations differ in morphological, physiological, and behavioral characteristics. Two populations that are morphologically very dissimilar are likely to be distinct species. It should be understood, however, that the level of morphological differentiation cannot be used with implicit confidence as a criterion of a species. For example, two species of fruit flies, *Drosophila pseudoobscura* and *Drosophila persimilis*, although reproductively isolated, are almost indistinguishable on morphological grounds.

Origin of Isolating Mechanisms

How do reproductive isolating mechanisms arise? In the 1940s, John A. Moore undertook a series of instructive evolutionary studies on the leopard frog, *Rana pipiens*. The leopard frog is widely distributed in North America, ranging from northern Canada through the United States and Mexico into the lower reaches of Central America. Moore obtained leopard frogs from different geographical populations and crossed them in the laboratory. When frogs from northeastern United States (Vermont) were crossed with their southerly distributed lowland relatives in eastern Mexico (Axtla in San Luis Potosi), the hybrid embryos failed to develop normally. Thus, the geographically extreme members of this species have diverged genetically to the extent that they cannot produce viable hybrids in the laboratory.

It must be admitted that the possibility of a Vermont frog crossing with a Mexican frog in nature is extremely remote. It took a biologist to bring these two frogs together. Yet, this emphasizes the point that an isolating mechanism, such as hybrid inviability, does not develop for the effect itself; it is simply the natural consequence of sufficient genetic differences having accumulated in two populations during a long geographical separation. The late Hermann J. Muller of Indiana University was among the first to advocate that isolating mechanisms originate as a by-product of genetic divergence of allopatric populations. The genetic changes that arise to adapt one population to particular environmental conditions may also be instrumental in reproductively isolating that population from other populations that are themselves developing adaptive gene complexes. Indeed, the embryos of Vermont leopard frogs differ considerably in their range of temperature tolerance from embryos of eastern Mexican frogs. It might well be, then, that the embryonic defects in hybrids between these northern and southern frogs are associated with the different temperature adaptations of the parental eggs.

If the Vermont and Mexican leopard frogs were ever to meet in nature, then any intercrosses between them would lead to the formation of inviable hybrids. This would represent a wastage of reproductive potential of the parental frogs. Theodosius Dobzhansky has advanced the interesting hypothesis that under such conditions, natural selection would promote the establishment of isolating mechanisms that would guard against the production of abnormal hybrids. In frogs, a normal mating or a mismating in a mixed population depends principally on the discrimination of the female. The productive potential is obviously lower for an undiscriminating female than for a female who leaves normal offspring. If the tendency to mismate is inheritable, then the genes responsible for this tendency will eventually be lost or sharply reduced in frequency by elimination of the indiscriminate females, an elimination effectively accomplished by the inviability of their offspring. Thus, the continual propagation of females that most resist the attentions of "foreign" males will lead eventually to a situation in which mismatings do not occur and abnormal hybrids are no longer produced.

Karl Koopman, an able student of Theodosius Dobzhansky, has tested the thesis that natural selection tends to strengthen, or make complete, the reproductive isolation between two species coexisting in the same territory. Koopman used for experimentation two species of fruit flies, *Drosophila pseudoobscura* and *Drosophila persimilis*. In nature, sexual selection between these two sympatric species is strong, and interspecific matings do not occur. However, in a mixed population in the laboratory, particularly at low temperatures, mismatings do take place. Koopman accordingly brought together members of both species in an experimental cage and purposely kept the case at a low temperature (16°C). Hybrid flies were produced and were viable, but Koopman in effect made them inviable by painstakingly removing them from the breeding case when each new generation emerged. Over a period of several generations the production of hybrid flies dwindled markedly and mismatings in the population cage were substantially curtailed. This is a dramatic demonstration of the efficacy of selection in strengthening reproductive isolation between two sympatric species.

Man: A Single Variable Species

There is only one present-day species of man, *Homo sapiens*. Different populations of man can interbreed successfully and, in fact, do. The extensive commingling of populations renders it difficult, if not impossible, to establish discrete racial categories in man. Races, as we have seen, are geographically defined aggregates of local populations. The populations of mankind are no longer sharply separated geographically from one another. Multiple migrations of peoples and innumerable intermarriages have tended to blur the genetical contrasts between populations. The boundaries of human races, if they can be delimited at all, are at best fuzzy, ever-shifting with time.

The term *race* is regrettably one of the most abused words in the English vocabulary. The biologist views a race as synonymous with a geographical subspecies; a race or subspecies is a genetically distinguishable subgrouping of a species. It is exceedingly important to recognize that a race is *not* a community based on language, literature, religion, nationality, or customs. There are Aryan languages, but there is no Aryan race. Aryans are peoples of diverse genetical makeups who speak a common tongue (Indo-European). *Aryan* is therefore nothing more than a linguistic designation. In like manner, there is a Jewish religion, but not a Jewish race. And there is an Italian nation, but not an Italian race. A race is a reproductive community of individuals occupying a definite region, and in one and the same geographical region may be found Aryans, Jews, and Italians. Every human population today consists of a multitude of diverse genotypes. A "pure" population or race, in which all members are genetically alike, is a myth and an absurdity.

Selected Readings

Dobzhansky, T. 1951. *Genetics and the origin of species.* New York: Columbia University Press.

Dowdeswell, W. H. 1960. *The mechanism of evolution.* New York: Harper & Row.

Goldsby, R. A. 1971. *Race and races.* New York: Macmillan Co.

King, J. C. 1971. *The biology of race.* New York: Harcourt Brace Jovanovich.

Mayr, E. 1942. *Systematics and the origin of species.* New York: Columbia University Press.

Mead, M., Dobzhansky, T., Tobach, E., and Light, R. E., eds. 1968. *Science and the concept of race.* New York: Columbia University Press.

Montagu, A. 1964. *The concept of race.* New York: Free Press of Glencoe.

Montagu, A. 1965. *Man's most dangerous myth: the fallacy of race.* New York: Macmillan Co.

Simpson, G. G. 1953. *The major features of evolution.* New York: Columbia University Press.

Stebbins, G. L. 1966. *Processes of organic evolution.* Englewood Cliffs, N.J.: Prentice-Hall.

Wallace, B., and Srb, A. 1964. *Adaptation.* Englewood Cliffs, N.J.: Prentice-Hall.

11

Cataclysmic Evolution

The process leading to the formation of a new species generally extends over a great reach of time. As we have seen, the origin of new species involves a long period of geographical isolation and the long-term influences of natural selection. A sudden and rapid emergence of a new kind of organism is scarcely imaginable. Yet, a natural mechanism does exist whereby a new species can arise rather abruptly. The process is associated with the phenomenon of *polyploidy,* or the multiplication of the chromosome complement of an organism. Species formation through polyploidy has occurred almost entirely, if not exclusively, in the plant kingdom. Many of our valuable cultivated crop plants, such as wheat, oats, cotton, tobacco, and sugar cane trace their origin to this cataclysmic or explosive type of evolution.

Wheat

The domestic wheats and their wild relatives have an intriguing evolutionary history. There are numerous species of wheat, all of which fall into three major categories on the basis of their chromosome numbers. The most ancient type is the small-grain einkorn wheat, containing 14 chromosomes in its body (somatic) cells. There are two species, one wild and the other cultivated, both of which may be found growing in the hilly regions of southeastern Europe and southwestern Asia. Cultivated einkorn has slightly larger kernels than the wild form, but the yields of each are low and the grain is used principally for feeding cattle and horses.

Another assemblage of wheat, once widely grown, is the emmer series, of which there are at least six species. The chromosome number in the nuclei of somatic cells of emmer wheats is 28. These varieties, found in Europe and the United States, are used today principally as stock feed, although one of them, called durum wheat, is of commercial value in the production of macaroni and spaghetti.

The most recently evolved, and by far the most valuable agriculturally, are the bread wheats. The bread wheats have not been known to occur in the wild state; all are cultivated types. The bread wheats have 42 chromosomes. These wheats, high in protein content, comprise almost 90 percent of all the wheat harvested in the world today.

The various species of wheats thus fall into three major groups, with 14, 28, and 42 chromosomes, respectively. A list of the representatives of these three groups is given in table 11.1.

Table 11.1

Species of Wheat (Triticum)

14 Chromosomes	28 Chromosomes	42 Chromosomes
T. aegilopoides (Wild Einkorn)	*T. dicoccoides* (Wild Emmer)	*T. aestivum* (Bread Wheat)
T. monococcum (Cultivated Einkorn)	*T. dicoccum* (Cultivated Emmer)	*T. sphaerococcum* (Shot Wheat)
	T. durum (Macaroni Wheat)	*T. compactum* (Club Wheat)
	T. persicum (Persian Wheat)	*T. spelta* (Spelt)
	T. turgidum (Rivet Wheat)	*T. macha* (Macha Wheat)
	T. polonicum (Polish Wheat)	

Origin of Wheat Species

Virtually all authorities are agreed on the sequence of evolutionary events depicted in figure 11.1. Einkorn wheat, possessing a chromosome number of 14, was doubtless one of the ancestral parents of the 28-chromosome emmer assemblage. A most remarkable, but generally accepted, thesis is that the other parent was not a wheat at all, but rather *Aegilops speltoides,* a wild grass with 14 chromosomes. This wild grass parent occurs as a common weed in the wheat fields of southwestern Asia. The cross of einkorn wheat and the wild grass would yield an F_I hybrid that possesses 14 chromosomes, 7 from each parent. We may designate the 7 chromosomes from one parent species as set (or genome) *A,* and the 7 from the other parent species as set (or genome) *B*. Accordingly, the F_I hybrids would have the *AB* genomes.

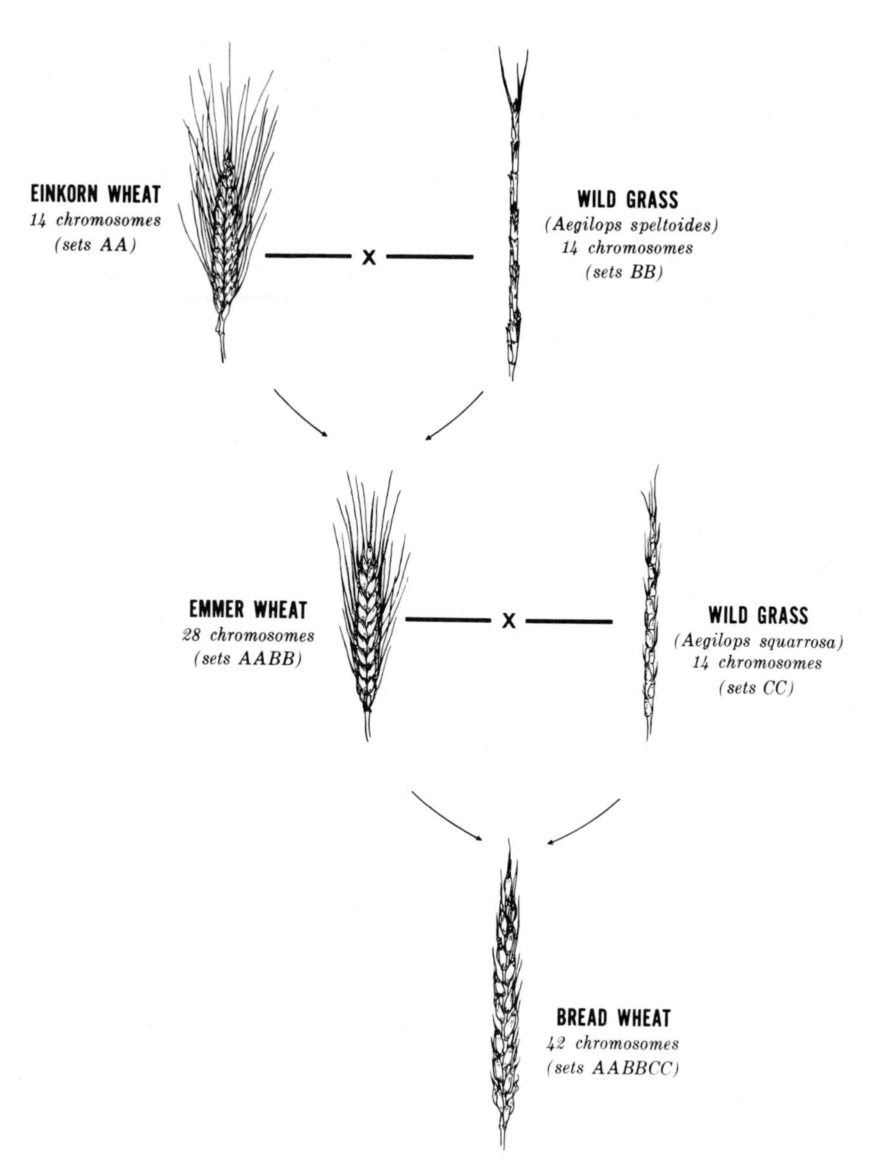

Figure 11.1 Evolution of wheat. Emmer wheat resulted from the hybridization of einkorn wheat with a wild grass, *Aegilops speltoides.* The common bread wheat is the product of hybridization of emmer wheat with a useless and noxious species of wild grass, *Aegilops squarrosa.* The hybridizations were followed by chromosome doubling, a phenomenon discussed in the text.

If the chromosome complement in the hybrid accidentally doubled, then the hybrid would contain 28 instead of 14 chromosomes and pass on the doubled set of chromosomes to its offspring. Such an event, strange as it may seem, accounts for the emergence of the 28-chromosome emmer wheat. This new species is characterized as having the *AABB* genomes.

In turn, the 28-chromosome emmer wheat was the ancestor of the 42-chromosome bread wheat. In the early 1900s, the British botanist John Percival hazarded the opinion that the bread wheat group arose by hybridization of a species of wheat of the emmer group (28 chromosomes) and goat grass, *Aegilops squarrosa,* a useless weed commonly found growing in wheat fields in the Mediterranean area. Although this startling suggestion was initially viewed with skepticism, it is currently conceded that Percival was correct. *Aegilops squarrosa* possesses 14 chromosomes, and thus would transmit 7 of its chromosomes (set *C*) to the hybrid. The hybrid would contain 21 chromosomes (sets *ABC*), having received 14 (sets *AB*) from its emmer wheat parent. The subsequent duplication in the hybrid of each chromosome set provided by the parents would result in a 42-chromosome wheat species (*AABBCC*).

The initial F_I hybrid between einkorn wheat and *Aegilops speltoides* (or between emmer wheat and *Aegilops squarrosa*) is sterile, but when the chromosome complement doubles, then a fully fertile species arises. Is it to be expected that the F_I hybrid would be sterile? And what would account for the fertility of the hybrid when chromosome doubling occurred? This requires a deeper look into the phenomenon of polyploidy, to which we shall now turn.

Mechanism of Speciation by Polyploidy

Figure 11.2 illustrates the underlying basis of the fertility of a formerly sterile hybrid resulting from the doubling of its chromosome number. For ease of presentation, the parental species are shown with a small number of chromosomes, 6 and 4, respectively. It should be noted that the chromosomes are present in pairs. The members of each pair are alike or homologous, but each pair is distinguishable from the other. Thus, the 6-chromosome parent possesses three different pairs of chromosomes; the 4-chromosome parent, two different pairs.

The gametes, egg cells and pollen cells (sperm), are derived by cell divisions of a special kind, called meiosis. One of the essential features of the process of meiosis is that the members of each pair are attracted to each other and come to lie side by side in the nucleus. The meiotic cell divisions are intricate, but the pertinent outcome is the separation of like, or homologous, chromosomes such that each gamete comes to possess only one member of each pair of chromosomes. Each gamete is said to contain a *haploid* complement of chromosomes, or half the number found in a somatic cell. The latter, in turn, is described as being *diploid* in chromosome number.

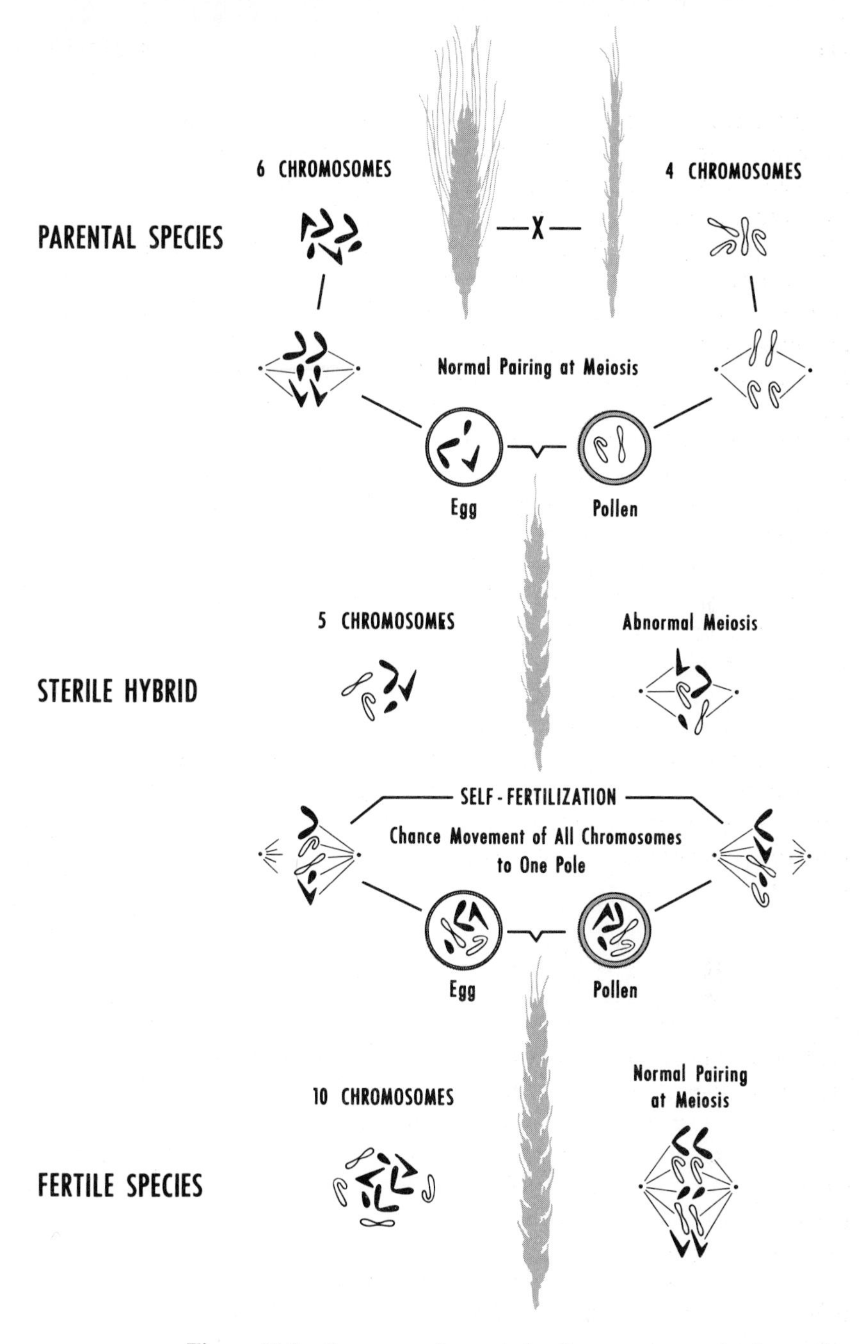

Figure 11.2 Sequence of events leading to a new, fertile species from two old species by hybridization and polyploidy. The F_1 hybrid plant derived from a cross of the two parental species is sterile. The F_1 hybrid may occasionally produce viable gametes when all chromosomes fortuitously enter a gamete during the process of meiosis. The fusion of such gametes leads to a new form of plant, which contains two complete sets of chromosomes (one full set of each of the original parents).

The sterility of the first-generation hybrid is now comprehensible. There are simply no homologous chromosomes in the F_I hybrid. Each chromosome lacks a homologue to act as its pairing partner at meiosis. The process of meiosis in the hybrid is chaotic; the chromosomes move at random into the gametes. The eggs and pollen cells typically contain an odd assortment of chromosomes and are nonfunctional.

Occasionally, by sheer chance, a few gametes might be produced by the F_I hybrid that contain all the chromosomes (fig. 11.2). These gametes would be functional, and the fusion of such sex cells would give rise to a plant that contains twice the number of chromosomes that the first-generation hybrid possessed. The plant actually would contain two complete sets of chromosomes; that is, the full diploid complement of chromosomes of each original parent. Such a double diploid is termed a polyploid—more specifically, a *tetraploid.*

The tetraploid hybrid would resemble the first-generation hybrid, but the plant as a whole would be larger and somewhat more robust as a consequence of the increased number of chromosomes. More importantly, the tetraploid hybrid would be fully fertile. The meiotic divisions would be normal, since each chromosome now has a regular pairing partner during meiosis (fig. 11.2). The tetraploid hybrid is a true breeding type; it can perpetuate itself indefinitely. It is, however, reproductively incompatible with its original parental species. If it were to cross with its original parental species, the offspring would be sterile. Hence, the tetraploid hybrid is truly a new distinct species. In but a few generations, we have witnessed essentially the fusion of two old species to form a single derived species.

Experimental Verification

To return to our wheat story, bread wheat (fig. 11.1) contains the chromosome sets of three diploid species—the *AA* of einkorn wheat, the *BB* of *Aegilops speltoides,* and the *CC* of *Aegilops squarrosa.* (The third set is typically referred to by botanists as the *D* genome. It is an accident of nomenclature that the third genome received the letter *D,* rather than *C.*) Technically, then, the bread wheat contains six sets of chromosomes; it is a *hexaploid.* The relationships of the three major groups of wheats can be shown as follows:

Einkorn = 14 = *AA* = diploid
Emmer = 28 = *AABB* = tetraploid
Bread = 42 = *AABBCC* = hexaploid

Experimental proof was lacking at the time John Percival proposed that the bread wheats originated from hybridization between the emmer wheat and goat grass, followed by chromosome doubling in the hybrid. Verification awaited an effective method of artificially inducing diploid cells to become polyploid. The search for an efficient chemical inducing agent culminated in the discovery in the late 1930s of colchicine, a substance obtained from the roots of the autumn crocus plant. Treatments of diploid plant cells with colchicine result in a high percentage of polyploid nuclei in the treated plant cells. Colchicine acts on the spindle apparatus of a dividing cell, and prevents a cell from dividing into two daughter halves. The treated undivided cell contains two sets of daughter chromosomes, which ordinarily would have separated from each other had cell division not been impeded. The cell thus comes to possess twice the usual number of chromosomes.

The experimental production of polyploid cells through the application of colchicine paved the way for studies on wheat by E. S. McFadden and E. Sears of the United States Department of Agriculture. These investigators successfully hybridized a tetraploid species of emmer wheat with the diploid wild grass *Aegilops squarrosa*. The chromosome number in the hybrid was doubled by treatment with colchicine. The synthetic hexaploid hybrids were similar in characteristics to natural hexaploid species of bread wheat, and produced functional gametes. At almost the same time, Hitoshi Kihara of Japan obtained a comparable hexaploid wheat species, which spontaneously and naturally had become converted from a sterile hybrid to a fertile hybrid. Kihara's work reinforced the notion that doubling of chromosomes can occur accidentally.

To complete the proof, McFadden and Sears crossed their artificially synthesized hexaploid wheat species with one of the naturally occurring bread wheat *Triticum spelta*. Fully fertile hybrids resulted, removing any doubt that wild grass, a noxious weed, is indeed a parental ancestor of the bread wheats.

Cotton

The story of the cataclysmic evolution of cotton (*Gossypium*) has been partially unraveled by botanical investigators. The evolution of cotton poses some interesting, but unresolved, problems.

Cotton is widely distributed throughout the world, and occurs in both the wild and the cultivated state. The cultivated types in the Americas are represented by *Gossypium barbadense,* the prominent cotton of South America, and *Gossypium hirsutum,* grown mainly in Central America and the United States. These two cultivated species of American cotton are of particular interest in that they each possess 52 chromosomes (or 26 pairs) in their cells. Thirteen of these pairs of chromosomes are small and resemble those in wild diploid cotton species still found growing in the Americas; the other 13 pairs are large and like those of diploid cotton species native to the Old World. The Old World cotton ancestor contains the *A* chromosome set (or genome); the wild American ancestor possesses the *D* genome. Unquestionably, the present-day American cultivated cotton resulted from a cross between the Old World cotton and the wild American cotton (*AA* x *DD*), with subsequent natural duplication of the chromosome complement in the hybrid (*AD* to *AADD*). The cultivated American cottons are thus tetraploid progenies of two diploid species. The question as to when and where the two diploid progenitors met and hybridized is a thorny one.

Available evidence indicates that the American tetraploid species arose by hybridization in the coastal valleys of Peru in western South America. Many botanists contend that the Old World diploid parent came from southern Asia, having reached South America by dispersal across the Pacific Ocean. The seeds of cotton, however, are not adapted to transportation over great distances by either water or wind. One authority on cotton, S. C. Harland, has suggested that the Old World parent crossed the Pacific by a land bridge in late Cretaceous times, about 100 million years ago. Most modern geologists dismiss as unreasonable the once popular notion of an ancient Pacific land bridge.

Other botanists, particularly J. B. Hutchinson, R. A. Silow, and S. G. Stephens, have argued that the Asiatic diploid cotton was introduced within historic times by civilized man. Early nomadic man from Asia carried the seeds of his crop plants across the Pacific to South America. From hybridizations of the transplanted Old World cotton with wild cotton of the valleys of Peru emerged the superior tetraploid plant. This tetraploid cotton was subsequently introduced into Central America, and then spread to the United States. The South American tetraploids are no longer reproductively compatible with their northern counterparts; hence, the presence today, as remarked earlier, of two distinct species of cultivated cotton in the Americas, *Gossypium barbadense* and *Gossypium hirsutum*.

The reconstruction of the past history of the cultivated American cotton is far from complete. There exists a primitive diploid species of cotton in south Africa, regarded as the forebear of the Old World diploid cottons. The Asiatic cottons were probably derived from the African type. Recently, there has been speculation that the African species of cotton, rather than the Asiatic species, was the immediate parent of the American cotton. This would be in accord with Wegener's theory of continental drift. In 1912, the Austrian meteorologist and Arctic explorer Alfred Wegener proposed that the earth's continents had once been a huge land mass (a supercontinent) and have reached their present geographical positions by splitting up and drifting across the ocean floors (fig. 11.3). Initially, Wegener's theory was ridiculed. At a meeting of the American Philosophical Society in Philadelphia in the 1920s, the key speaker pronounced Wegener's thesis "utter, damned rot." Today, the idea of continental drift has been revived and has gained a high measure of scientific respectability. Indeed, we now recognize that Wegener was essentially correct.

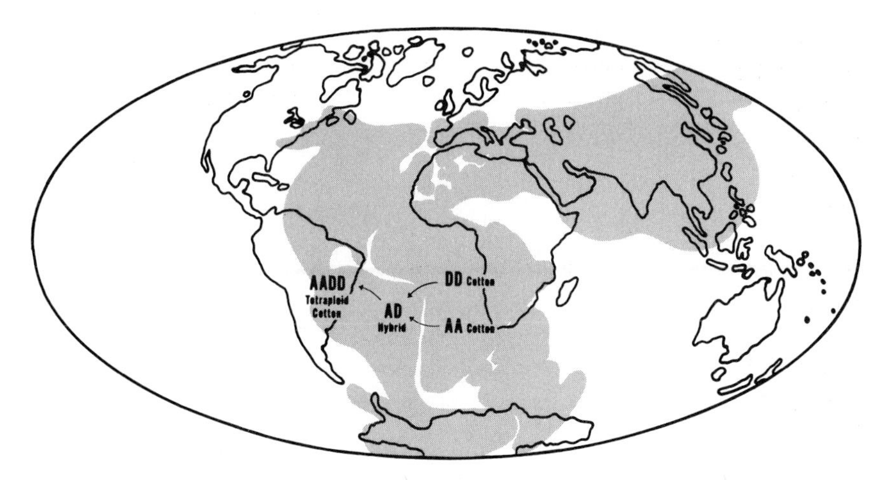

Figure 11.3 Wegener's theory of continental drift. The present continents were joined in a single land mass (*Pangaea*) before the start of the Mesozoic era (about 200 million years ago). The hybridization of two African diploid species of cotton to give rise to the American tetraploid species of cotton may have occurred before the continents drifted apart.

Cotton and Continental Drift

The original single land mass, called *Pangaea,* split and began to move apart about 180 to 200 million years ago. The supercontinent first split into two blocks: *Laurasia* in the Northern Hemisphere, consisting of North America, Europe, and Asia; and *Gondwanaland* in the Southern Hemisphere, made up of Africa, South America, Antarctica, Australia, and India (which then lay far south of Asia). South America began sliding westward about 135 million years ago when a rift appeared between South America and Africa. Antarctica parted from Africa, and India broke free to move northerly toward Asia. North America separated from Europe only about 80 million years ago. It is of interest that India collided with Asia less than 40 million years ago. The collision uplifted the high Tibetan Plateau and raised the Himalayas.

The concept of continental drift serves to explain the present similarity in certain animals and plants in regions of Africa and South America. With respect to cotton, consideration may be given to the possibility that an ancestral diploid cotton species of south African (genome *A*) at one time overlapped the distribution of another ancestral species (genome *D*) in that part of western Africa adjacent to the American continent before continental drift (fig. 11.3). In the area of contact, hybridization between the two diploid species may have occurred naturally, giving rise to the American tetraploid species that became isolated in northern South America at the time of continental separation. This hypothesis, although provocative, limps a little. Africa and South America began to separate 135 million years ago, and perhaps earlier. The fossil history of flowering plants does *not* extend into the geologic past farther back than the Cretaceous period, about 135 million years ago (see chapter 14). It seems unlikely that any of the cotton species could have been in existence prior to the separation of the continents when there is essentially no fossil record of flowering plants before the Cretaceous. The hypothesis could be salvaged by assuming that there was a long pre-Cretaceous evolution of flowering plants, but that pre-Cretaceous conditions were not favorable for the preservation of their fossils.

Selected Readings

Briggs, D. and Walters, S. M. 1969. *Plant variation and evolution.* New York: McGraw-Hill Book Co.

Curtis, B. C. and Johnston, D. R. 1969. Hybrid wheat. *Scientific American,* May, pp. 21–29.

Hurley, P. M. 1968. The confirmation of continental drift. *Scientific American,* April, pp. 53–64.

Kurten, B. 1969. Continental drift and evolution. *Scientific American,* March, pp. 54–64.

Mangelsdorf, P. C. 1953. Wheat. *Scientific American,* July, pp. 50–59.

Stebbins, G. L. 1950. *Variation and evolution in plants.* New York: Columbia University Press.

Stebbins, G. L. 1951. Cataclysmic evolution. *Scientific American,* April, pp. 54–59.

12 Adaptive Radiation

The capacity of a population of organisms to increase its numbers is largely governed by the availability of resources—food, shelter, and space. The available supply of resources in a given environment is limited, whereas the organism's innate ability to multiply is unlimited. A particular environment will soon prove to be inadequate for the number of individuals present. It might thus be expected that some individuals would explore new environments where competition for resources is low. The tendency of individuals to exploit new opportunities is a factor of major significance in the emergence of several new species from an ancestral stock. The successful colonization of previously unoccupied habitats can lead to a rich multitude of diverse species, each better fitted to survive and reproduce under the new conditions than in the ancestral habitat. The spreading of populations into different environments accompanied by divergent adaptive changes of the emigrant populations is called *adaptive radiation.*

Galápagos Islands

One of the biologically strangest, yet fascinating, areas of the world is an isolated cluster of islands of volcanic origin in the eastern Pacific, the Galápagos Islands. These islands, which Darwin visited for five weeks in 1835, lie on the equator, 600 miles west of Ecuador (fig. 12.1). The islands are composed wholly of volcanic rock; they were never connected with the mainland of South America. The rugged shoreline cliffs are of gray lava and the coastal lowlands are parched, covered with cacti and thorn brushes. In the humid uplands, tall trees flourish in rich black soil.

Giant land-dwelling tortoises still inhabit these islands. After many years of being needlessly slain by pirates and whalers, these remarkable

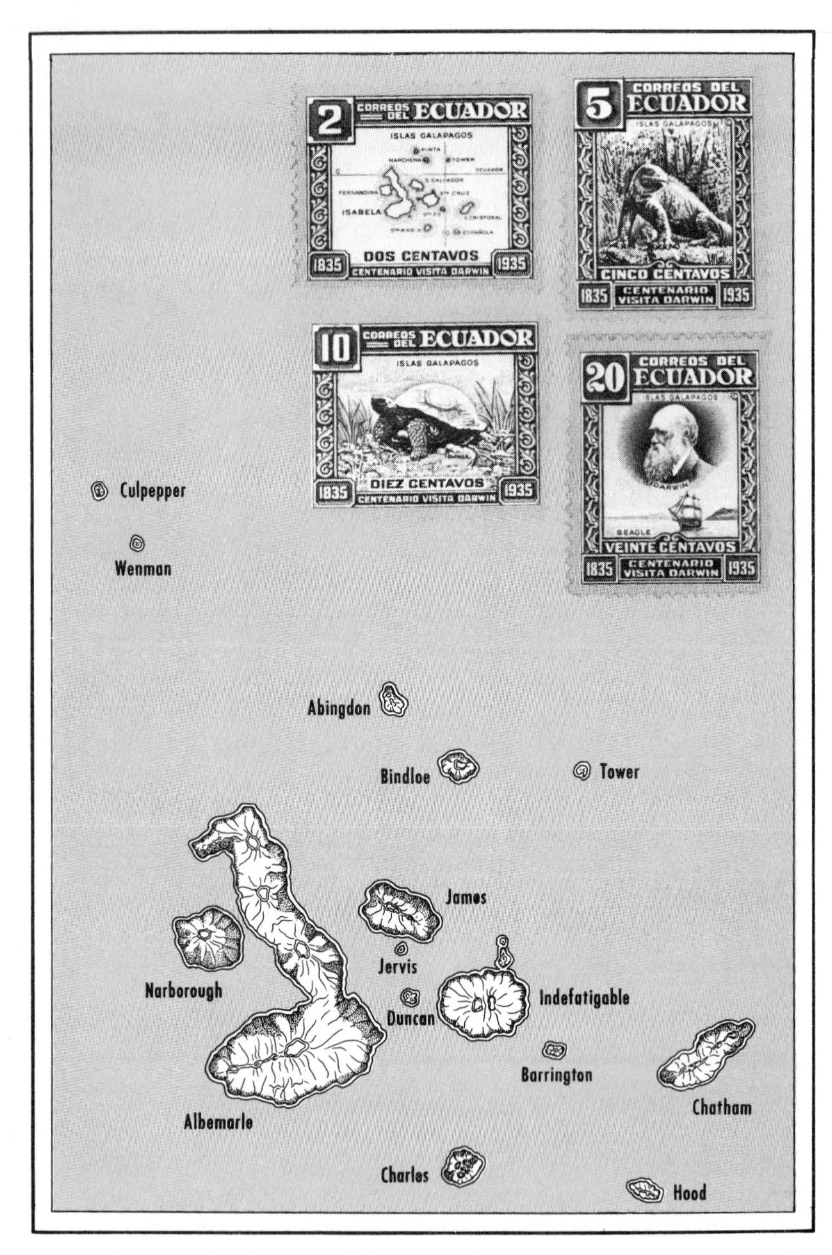

Figure 12.1 Galápagos Islands ("Enchanted Isles") in the Pacific Ocean, 600 miles west of Ecuador. Darwin had explored this cluster of isolated islands, and found a strange animal life, a "little world within itself." The four stamps shown were issued by Ecuador to commemorate the centenary of Darwin's visit in 1835.

animals now live protected in a sanctuary created in 1959. Still prevalent on the islands are the world's only marine iguana and its inland variety, the land iguana (fig. 12.2). These two species of prehistoric-looking lizards are ancient arrivals from the mainland. The marine forms occur in colonies on the lava shores, and swim offshore to feed on seaweed. The land iguana lives on leaves and cactus plants. Cactus fills most of the water needs of the land iguana.

At least 85 different kinds of birds have been recorded on the islands. These include rare cormorants that cannot fly, found only on Narborough Island, and flamingos, which breed on James Island. Of particular interest are the small black finches. These black birds exhibit remarkable variations in the structure of the beak and in feeding habits. The finches afford an outstanding example of adaptive radiation. It was the marked diversity within this small group of birds that gave impetus to Darwin's evolutionary views. Darwin had correctly surmised that the diverse finches were modified descendants of the early, rather homogeneous, colonists of the Galápagos. Our present knowledge of these birds, now appropriately called "Darwin's

Figure 12.2 Land Iguana on one of the Galápagos Islands. Despite their horrendous appearance, these bizarre inland lizards are mild, torpid, and vegetarians. They feed on leaves and cactus plants. (Courtesy of the American Museum of Natural History.)

finches," derives largely from the accomplished work of David Lack at Oxford, who visited the Enchanted Isles in 1938.

Darwin's Finches

Darwin's finches descended from seed-eating birds that inhabited the mainland of South America. The ancestors of Darwin's finches were early migrants to the Galápagos Islands, and probably the first land birds to reach the islands. These early colonists have given rise to 14 distinct species, each well adapted to a specific niche (fig. 12.3). Thirteen of these species occur in the Galápagos; one is found in the small isolated Cocos Island, northeast of the Galápagos.

The most striking differences among the species are in the sizes and shapes of the beak, which are correlated with marked differences in feeding habits. Six of the species are ground finches, with heavy beaks specialized for crushing seeds. Some of the ground finches live mainly on a diet of seeds found on the ground; others feed primarily on the flowers of prickly pear cacti. The cactus eaters possess decurved, flower-probing beaks. Their beaks are thicker than those of typical flower-eating birds.

All the other species are tree finches, the majority of which feed on insects in the moist forests. One of the most remarkable of these tree dwellers is the woodpecker finch. It possesses a stout, straight beak, but lacks the long tongue characteristic of the true woodpecker. Like a woodpecker, it bores into wood in search of insect larvae, but then it uses a cactus spine or twig to probe out its insect prey from the excavated crevice. Equally extraordinary is the warbler finch, which resembles in form and habit the true warbler. Its slender, warblerlike beak is adapted for picking small insects off bushes. Occasionally, like a warbler, it can capture an insect in flight.

Factors in Diversification

No such great diversity of finches can be found on the South American mainland. In the absence of vacant habitats on the continent, the occasion was lacking for the mainland birds to exploit new situations. However, given the unoccupied habitats on the Galápagos Islands, this was opportunity for the invading birds to evolve in new directions. In the absence of competition, the colonists occupied several ecological habitats, the dry lowlands as well as the humid uplands. The finches adopted modes of life that ordinarily would not have been opened to them. If true warblers and true woodpeckers had already occupied the islands, it is doubtful that the finches could have evolved into warblerlike and woodpeckerlike forms. Thus, *a prime factor promoting adaptive radiation is the absence of competition.*

The emigration of the ancestral finches from the mainland was assuredly not conscious or self-directed. The dispersal of birds from the original home was at random, resulting from the pressure of increasing numbers on the

Figure 12.3 Representatives of Darwin's finches. There are 14 species of Darwin's finches, confined to the Galápagos with the exception of one species that inhabits Cocos Island. Closest to the ancestral stock are the six species of ground finches, primarily seed-eaters. The others evolved into eight species of tree finches, the majority of which feed on insects.

means of subsistence. By chance, some of the birds reached the Galápagos Islands. The original flock of birds that fortuitously arrived at the islands was but a small sample of the parental population, containing at best a limited portion of the parental gene pool. It may be that only a small amount of genetic variation was initially available for selection to work on. What evolutionary changes occurred at the outset were mainly due to random survival (genetic drift). However, the chance element would become less important as the population increased in size. Selection unquestionably became the main evolutionary agent, molding the individual populations into new shapes by the preservation of new favorable mutant or recombination types. More than one island was colonized, and the complete separation of the islands from each other promoted the genetic differentiation of each new local population.

Competitive Exclusion

When two populations of different species are obliged (under experimental conditions) to use a common nutrient, the two will compete with each other for the common resource. Under competition for the identical needs, only one of the two species populations will survive; the other will be eliminated or excluded by competition. This is the principle of *competitive exclusion*. It is also known as *Gause's principle*, after the Russian biologist G. F. Gause, who first demonstrated experimentally that under controlled laboratory conditions, one of the two competing species perishes.

In 1934, Gause studied the interactions under carefully controlled culture conditions of two protozoan species, *Paramecium caudatum* and *Paramecium aurelia*. When each species was grown separately in a standard medium in a test tube containing a fixed amount of bacterial food, each species flourished independently. When the two species were placed together in the same culture vessel, however, the growth of *P. caudatum* gradually diminished until the population became eliminated. In enforced competition for the same limited food supply, *P. aurelia* was the more successful species.

Gause enunciated his principle primarily on the basis of observations of "bottle populations" in the laboratory. The competition experiments reveal that two species populations cannot exist together if they are competing for precisely the same limited resource. Alternatively, if two species in nature were to occupy the same habitat, the expectation is that each would have different ecological requirements, even though the degree of difference is slight. Ecologists have demonstrated the validity of this view. In virtually every natural situation carefully examined, two co-inhabiting species have been found to differ in some requirement. The heterogeneous resources of the environment in a given locality are typically partitioned among the co-inhabiting species to minimize direct competition and enable the two (or more) species to coexist.

Most of the Galápagos Islands are occupied by more than one species of finch. On islands where several species of finches exist together, we find that each species is adapted to a different ecological niche. The three common species of ground finches—small (*Geospiza fuliginosa*), medium (*Geospiza fortis*), and large (*Geospiza magnirostris*)—occur together in the coastal lowlands of several islands. Each species, however, is specialized in feeding on a seed of a certain size. The small-beaked *Geospiza fuliginosa*, for example, feeds on small grass seeds, whereas the large-beaked *Geospiza magnirostris* eats large, hard fruits. Different species, with different food requirements, can thus exist together in an environment with varied food resources.

Coexistence

Gause's principle may be stated in the following form: No two species with identical requirements can continue to exist together. However, it is exceedingly unlikely that two species in nature would have *exactly* the same requirements for food and habitat. The sum total of environmental requirements for a species to thrive and reproduce has been termed the *niche* of that species population. The term *niche,* as ecologists use it, is more than simply the physical space that the species population occupies. It is essentially the way of life peculiar to a given species: its structural adaptations, physiological responses, and behavior within its habitat. Experience has shown that the likelihood of two species having identical niches is almost nil.

Direct evidence of the process of competition between species in nature is difficult to obtain. Observable competitive interaction is a relatively fleeting stage in the relation of two species populations. What the ecologist observes is the end result of competitive contact, when the actual or potential competitors have become differentially specialized to exploit different components of a local environment and accordingly live side by side. The outcome, then, of incipient competitive interaction is the *avoidance* of competition through differential specialization, or in the terminology of ecologists, through *niche diversification.*

Indirect arguments have been used to support the view that two closely related species populations come to exploit different ecological niches in the same locality after a beginning of competitive interaction. We may envision a situation in which two closely related species, with almost similar ecological requirements, expand their geographic ranges and meet in a common habitat. It may be presumed that each of the two species populations has appreciable genetic variability, and that the resources in the common habitat are varied. The two species populations would initially compete for suitable ecological niches in the new common habitat. However, we can expect that the members of the two species will ultimately become so different in structure and behavior that each species will become specialized to use different components of the environment. In other words, if genetic differences in morphological and behavioral characteristics tend at first to reduce competition between the two species, then subsequently natural selection will act to augment the differences between the two competing species. It is especially noteworthy that the differences between the two species become more pronounced as a consequence of selection *reducing* competition rather than *intensifying* competition.

A striking case of niche diversification has been described by David Lack for certain species of *Geospiza* in the Galápagos. Where two species occur together on an island, there is a conspicuous difference in the size of the beaks and food habits. Where either species exists alone on other islands, the beak is adapted to exploit more than one food resource. Thus, on the island of Tower (fig. 12.4), the cactus-feeding *Geospiza conirostris* and the

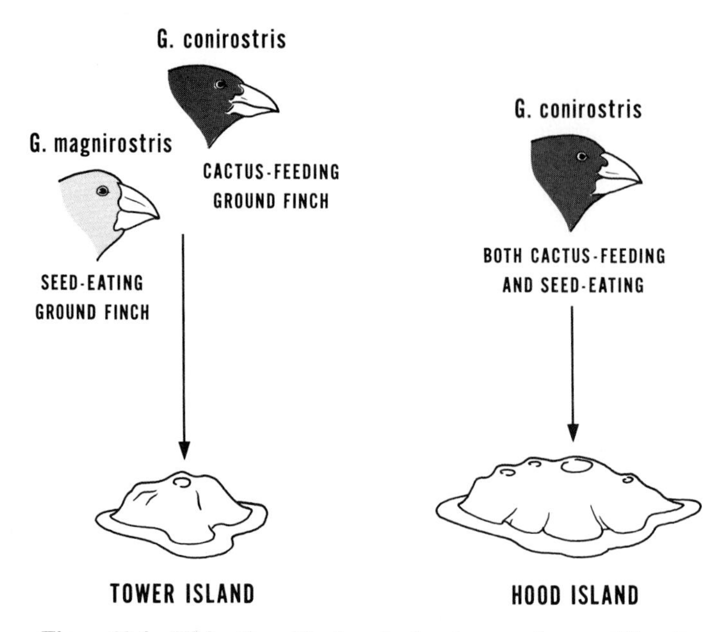

Figure 12.4 Niche diversification. In the absence of competition, the ground finch *Geospiza conirostris* on Hood Island has evolved a large beak that enables it to exploit a variety of food items (seed and cacti). On the island of Tower, the same ground finch *Geospiza conirostris* has evolved a specialized beak adapted solely for cactus feeding, thus reducing or avoiding competition for food resources with the seed-eating ground finch *Geospiza magnirostris*. The latter species does not exist on Hood Island.

large seed-eating ground finch *Geospiza magnirostris* live side by side. The former species occurs also on the island of Hood, but the latter species is absent, presumably having failed to invade or reach the island. In the absence of competition, *Geospiza conirostris* on Hood Island has evolved a larger beak that is adapted to feeding on both cactus and seeds. The sharp separation in beak size of the two species on Tower Island is understandable if we assume that competition initially fostered the differentiation of the beaks to permit each of the two species to adapt to a limited or restricted range of the available food resources. Each species is not genetically specialized in food habits, and competition between the two is now avoided.

There are many examples to illustrate how two or more species avoid competition. Among warblers that inhabit the spruce forests of Maine, each species confines its feeding to a particular region of the spruce tree. The myrtle warbler, for example, preys on insects in and below the spruce tree, whereas the Blackburnian warbler prefers those insects on the exterior leaves of the top of the tree. They can exist together because they use different resources of the same tree.

We may conclude by stating, paradoxically, that competition between two species populations achieves the avoidance or reduction of further competition, and not an intensification. In natural populations, coexistence of two species, rather than competitive exclusion, is the general rule.

Geographic Speciation

In an earlier chapter we said that geographic or spatial isolation is an important, if not essential, condition for the formation of a new species. When two populations are separated by a geographical barrier, each may diverge independently of the other. The distribution of insectivorous tree finches on some of the Galápagos Islands attests to the significance of geographical isolation in the speciation process.

One particular assemblage of tree finches has a widespread distribution (fig. 12.5). These finches now show a discontinuous pattern of geographical variation. If we were able to gaze into the evolutionary past, we would probably witness the series of events depicted in figure 12.5. At one time, we would be able to recognize three well-defined geographical groups, which, for simplicity of presentation, may be designated *A*, *B*, and *C*. We might then subsequently observe the Albemarle form (*C*) spreading to Charles Island and establishing a colony, which, in time, differentiated into group *C′*. The Albemarle finch today does show a close resemblance to the Charles form, which suggests that the latter was derived from the former. If we were to stop the evolutionary clock at this time, we would properly regard this assemblage of finches as a single species, subdivided into four well-marked geographic races. In fact, this assemblage has been designated as one species, *Camarhynchus psittacula*.

At the present time, we find that Charles Island is inhabited not only by group *C′* but also by a newly established colony, whose members are indistinguishable from those of group *A*. Undoubtedly, the *A* group on Charles Island comprises recent arrivals from one of the central islands, Indefatigable or Barrington.

The more recent immigrants, group *A*, on Charles occur together with the older colonists, group *C′*, but they do *not* interbreed. Thus, the two groups, having arisen initially from separate islands, had developed sufficient genetic differences in isolation to remain distinct when they met on Charles Island. Groups *A* and *C′* are two separate species, actually designated as *Camarhynchus psittacula* and *Camarhynchus pauper*, respectively.

This situation raises some interesting questions. If group *C* on Albemarle

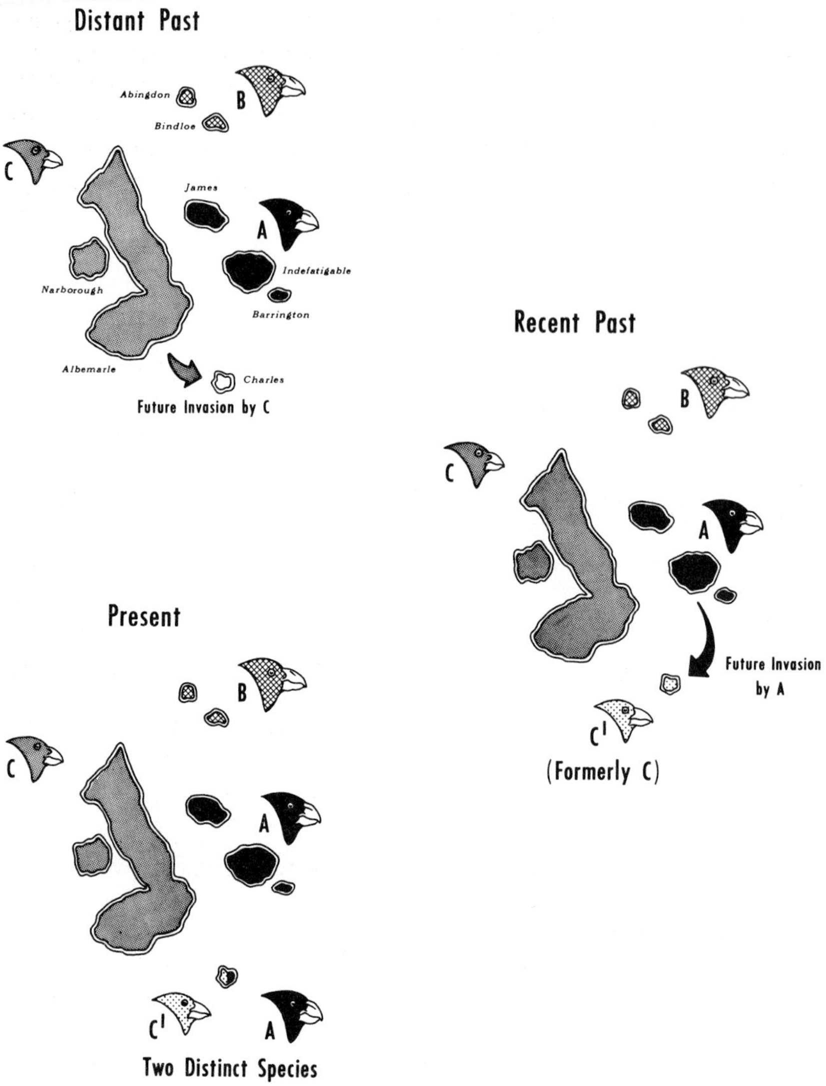

Distant Past

Abingdon
B
Bindloe
C
James
A
Indefatigable
Narborough
Barrington
Albemarle
Charles
Future Invasion by C

Recent Past

B
C
A
Future Invasion by A
C'
(Formerly C)

Present

B
C
A
C'
A
Two Distinct Species

Figure 12.5 Historical events leading to the existence today of two distinct species of tree finches, *Camarhynchus psittacula* (designated *A*) and *Camarhynchus pauper* (*C'*), which occur together on Charles Island. Charles Island had been colonized by finches on two separate occasions. The earlier immigrant came from Albemarle Island in the distant past; the later immigrant was a more recent arrival from Indefatigable or Barrington. The two immigrating groups, although initially only races of the same species, had accumulated sufficient genetic differences in isolation that they did not interbreed when they met on Charles Island. Thus, they are separate species. (Based on the extensive studies by David Lack.)

were now to spread to Abingdon, where *B* occurs, and an exchange of genes did not take place between them, then *B* and *C* would be considered as distinct species. But, as far as we know, *C* has not invaded the territory of *B*, and the proper designations of the two groups can not be made. Most evolutionists would probably regard *B* and *C* as races, or subspecies, of the same species, *Camarhynchus psittacula*. Regardless of the correct terminology, the important consideration is the demonstration that species formation is preceded by geographical separation.

Classification

The original ancestral stock of finches on the Galápagos diverged along several different paths. The pattern of divergence is reflected in the biologists' scheme of classification of organisms. All the finches are related to one another, but the various species of ground finches obviously are more related by descent to one another than to the members of the tree-finch assemblage. As a measure of evolutionary affinities, the ground finches are grouped together in one genus (*Geospiza*) and the tree finches are clustered in another genus (*Camarhynchus*). The different lineages of finches are portrayed in figure 12.6. It should be clear that our classification scheme is nothing more than an expression of evolutionary relationships between groups of organisms.

In the next chapter we shall see how adaptive radiation on a much larger scale than that which occurred in the finches led to the origin of radically new assemblages of organisms, distinguishable as orders and classes by the taxonomist.

Selected Readings

Dillon, L. S. 1973. *Evolution: concepts and consequences.* Saint Louis: C. V. Mosby Co.

Grant, V. 1963. *The origin of adaptations.* New York: Columbia University Press.

Hamilton, T. H. 1967. *Process and pattern in evolution.* New York: Macmillan Co.

Lack, D. 1947. *Darwin's finches.* London: Cambridge University Press.

Lack, D. 1953. Darwin's finches. *Scientific American,* April, pp. 66–72.

Mayr, E. 1963. *Animal species and evolution.* Cambridge, Mass.: Harvard University Press.

Merrell, D. J. 1962. *Evolution and genetics.* New York: Holt, Rinehart and Winston.

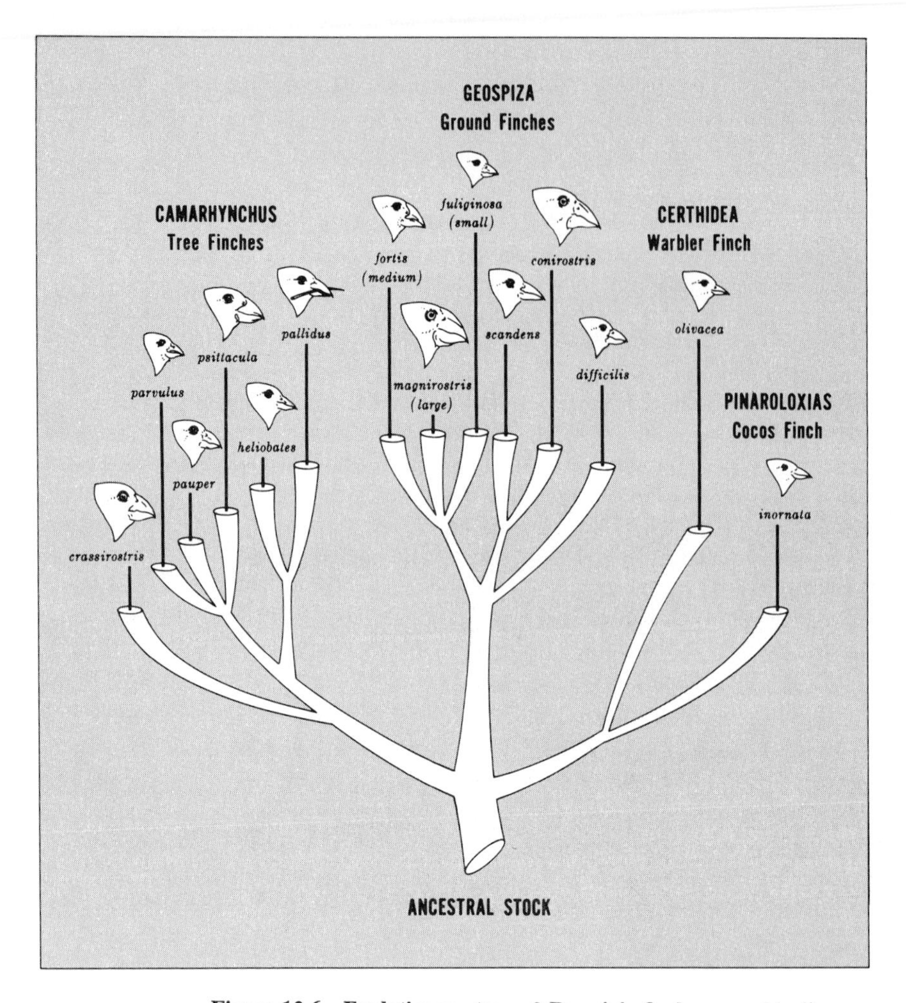

Figure 12.6 Evolutionary tree of Darwin's finches, graphically expressing what is known or surmised as to the degree of relationship or kinship between the different species of finches. Darwin's finches evolved from a common stock, and are represented today by 14 species. The species are assembled in four genera, of which two (Certhidea and Pinaroloxias) contain only a single species each. The species associated together in a genus have significant common attributes, judged to denote evolutionary affinities. (Based on the findings of David Lack.)

Moody, P. A. 1962. *Introduction to evolution.* New York: Harper & Row.

Moore, R., and editors of Time-Life Books. 1968. *Evolution.* New York: Time-Life Books.

Smith, J. M. 1958. *The theory of evolution.* Baltimore: Penguin Books.

13

Major Adaptive Radiations

The diversity of Darwin's finches had its beginning when migrants from the mainland successfully invaded the variety of vacant habitats on the Galápagos Islands. The pattern of adaptive radiation manifested by Darwin's finches has been imitated repeatedly by different forms of life. Organisms throughout the ages have seized new opportunities open to them by the absence of competitors and have diverged in the new environments. The habitats available to Darwin's finches were certainly few in comparison with the enormous range of ecological habitats in the world. The larger the region and the more diverse the environmental conditions, the greater the variety of life.

Approximately 400 million years ago, during a period of history that geologists call the Devonian, the vast areas of land were monotonously barren of animal life. Save for rare creatures like scorpions and millipedes, animal life of those distant years was confined to the water. The seas were crowded with invertebrate animals of varied kinds. The fresh and salt waters contained a highly diversified and abundant assemblage of cartilaginous and bony fishes. The vacant terrestrial regions were not to remain long unoccupied. From one of the many groups of fishes inhabiting the pools and swamps in the Devonian period emerged the first land vertebrate. The initial modest step onto land started the vertebrates on their conquest of all available terrestrial habitats. The story of the origin and diversification of the backboned land dwellers began then and has continued.

Invasion of Land

Prominent among the numerous Devonian aquatic forms were the lobe-finned fishes, the Crossopterygii, which possessed the ability to gulp air when they rose to the surface. These ancient air-breathing fishes represent the stock from which the first land vertebrates, the amphibians, were derived (fig. 13.1). The factors that led these ancestral lobe-finned fishes to venture onto land are unknown. The impelling force might have been population pressure or simply the inherent tendency of individuals, particularly of the young, to disperse. A. S. Romer of Harvard University has suggested that the crossopterygians were forced to crawl on dry land on those occasions when the pools they inhabited became foul, stagnant, or completely dry. There is convincing geological evidence that the Devonian years were marked by excessive seasonal droughts. It is not unimaginable that the water in some pools periodically evaporated. The suggestion, then, is that the crossoptery-

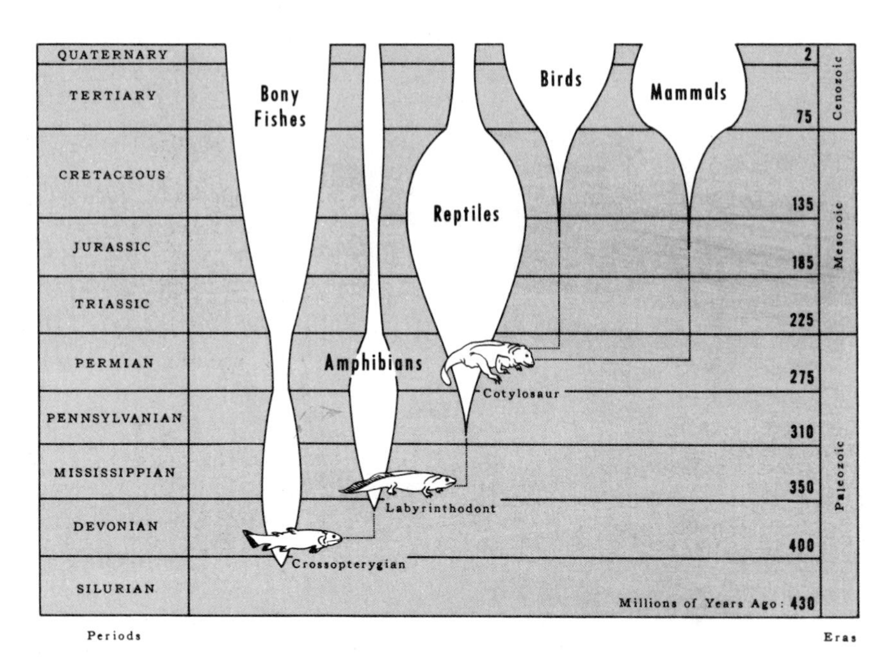

Figure 13.1 Evolution of land vertebrates in the geologic past. From air-breathing, lobe-finned fishes (crossopterygians) emerged the first four-footed land inhabitants, the amphibians. Primitive amphibians (labyrinthodonts) gave rise to the reptiles, the first vertebrates to become firmly established on land. The birds and mammals owe their origin to an early reptilian stock (cotylosaurs). An important biological principle reveals itself: each new vertebrate group did not arise from highly developed or advanced members of the ancestral group, but rather from early primitive forms near the base of the ancestral stock. The thickness of the various branches provides a rough measure of the comparative abundance of the five vertebrate groups during geologic history. The Devonian period is often called the *Age of Fishes;* the Mississippian and Pennsylvanian periods (frequently lumped together as the Carboniferous period) are referred to as the *Age of Amphibians;* the Mesozoic era is the grand *Age of Reptiles;* and the Cenozoic era is the *Age of Mammals.*

gians wriggled out of stagnant and shrinking water holes onto land to seek pools elsewhere in which water still remained. Thus, paradoxically, the first actual movements on land might not have been at all associated with an attempt to abandon aquatic existence, but rather to retain it.

The above hypothesis is admittedly speculative. However, the fact remains that these crossopterygians that emerged on land, though crudely adapted for terrestrial existence, did not encounter any competitors that could immediately spell doom to their awkward initial trial on land. We should note that the lobe-finned fishes did possess certain capacities that would prove to be important under the new conditions of life. Evolutionists speak of such potential adaptive characters as *preadaptations*.

The preadaptations of the lobe-finned fishes included primitive membranous lungs and internal nostrils, both of which are important for atmospheric breathing. It should be understood that such preadapted characters were not favorably selected with a view to their possible use in some future mode of life. There is no foresight or design in the selection process. Nor do mutational changes occur in anticipation of some new environmental condition. A trait is selected only when it imparts an advantage to the organism in its immediate environment. Accordingly, lungs in the crossopterygians did not evolve with conscious reference toward a possible future land life, but only because such a structure was important, if not essential, to the survival of these air-breathing fishes in their immediate surroundings.

The crossopterygians did not, of course, possess typical amphibian limbs. However, their lateral fins contained fleshy lobes, within which were bony elements that were basically comparable to those of a limb of a terrestrial vertebrate. Figure 13.2 shows a restoration of a widespread Devonian form, *Eusthenopteron*, in which the lateral fins had developed into stout, muscular paddles.

Before the close of the Devonian period, the transition from fish to amphibian had been completed. The early land-living amphibians were slim-bodied with fishlike tails, but having limbs capable of locomotion on land. The four-footed amphibians flourished in the humid swamps of Mississippian and Pennsylvanian times, but never did become completely adapted for existence on land. All the ancient amphibians, such as *Diplovertebron* (fig. 13.2), spent much of their lives in water, and their modern descendants—the salamanders, newts, frogs, and toads—must return to water to deposit their eggs. Thus, the amphibians were the first vertebrates to colonize land, but were, and still are, only partially adapted for terrestrial life.

Conquest of Land

From the amphibians emerged the reptiles, true terrestrial forms. The appearance of a shell-covered egg, which can be laid on land, freed the reptile from dependence on water. The elimination of a water-dwelling stage was a signif-

Figure 13.2 Stages in the transition of the lobe-finned fishes into amphibians, as reconstructed by W. K. Gregory and painted by Francis Lee Jaques. *Bottom,* the primitive Devonian air-breathing crossopterygian, *Eusthenopteron,* floundering on a stream bank with its muscular, paddle-like fins. *Top,* the Pennsylvanian tailed amphibian, *Diplovertebron,* with limbs capable of true locomotion on land. Much of the life of this early tetrapod was spent in water. (Courtesy of the American Museum of Natural History.)

icant evolutionary advance. The first primitive reptile most likely arose during Carboniferous times, but the fossil beds of this period have yet to reveal the appropriate reptilian ancestor. The ancestral reptile probably possessed the body proportions and well-developed limbs of the more advanced terrestrial forms of amphibians, such as *Seymouria* (fig. 13.3). *Seymouria* is not a reptile, but its skeletal features suggest terrestrial habits. This advanced amphibian may have been a descendant of an earlier amphibian group in the lower Carboniferous that was ancestral to the reptiles.

A key feature of reptiles (and higher vertebrates) is the *amniotic egg* (fig. 13.4). The egg of a reptile (or bird) contains a large amount of nourishing yolk. Moreover, development of the embryo takes place entirely within a thick shell. But these circumstances call for a special provision whereby

Figure 13.3 Seymouria, an advanced amphibian with reptilian-like body proportions (short trunk region and well-developed limbs). *Seymouria* is known only from Permian rocks (275 million years ago), long after the reptiles had appeared. It may represent a relict of an earlier amphibian group that was ancestral to the reptiles. (From a painting by Francis Lee Jaques; courtesy of the American Museum of Natural History.)

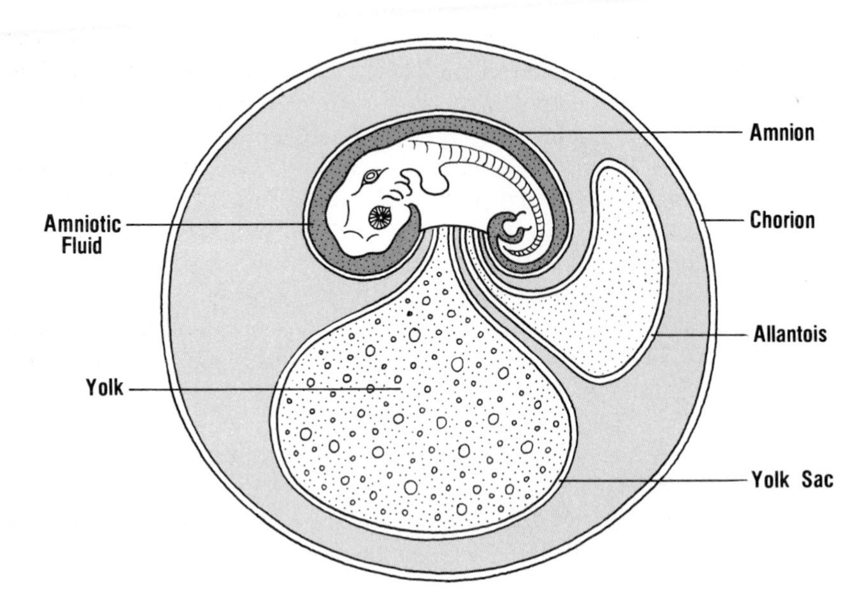

Figure 13.4 The amniotic egg. The developing embryo is enclosed by the membranous *amnion* and cushioned by amniotic fluid. The large reserve supply of food (yolk) is contained within the *yolk sac*. The sacklike *allantois* serves as a receptacle for the embryo's waste products. The outermost, vascularized enveloping membrane is the *chorion*.

food derived from the yolk and oxygen obtained from an external source can be made accessible to all parts of the developing embryo. The embryo itself constructs a complex system of membranes, known as *extraembryonic membranes*, which serve for protection, nutrition, and respiration. These are the *amnion, chorion, yolk sac,* and *allantois*.

When fully developed, the amnion is a thin membrane loosely enclosing the embryo. It does not fit the embryo snugly. The space between the amnion and the embryo, or amniotic cavity, is filled with a watery fluid, the amniotic fluid. The fluid is a cushion that protects the embryo from mechanical pressures and impacts and, at the same time, allows it freedom of movement. The chemical composition of the amniotic fluid resembles the chemical makeup of blood. Thus, the embryo during its development is bathed by a fluid that is compatible in its chemical nature with the embryo's blood. An interesting fact is that for the embryos of birds and reptiles, as well as for the human fetus, the amniotic fluid resembles seawater in chemical composition, in terms of elements (ions) such as sodium, potassium, calcium, and magnesium. After approximately 400 million years of evolution on land, vertebrates still carry essentially the chemical composition of seawater in their fluids—amniotic fluid, blood, and other body fluids. This has been interpreted as indicating that life originated in the sea, and that the balance of salts in various body fluids did not change very much in subsequent evolution. The amnion has been picturesquely characterized as a sort of private aquarium in which the embryo of land-living vertebrates recapitulate the water-living mode of existence of their remote ancestors.

As figure 13.4 shows, the enormous mass of yolk has been enclosed by an internal circular membrane, the yolk sac. This membrane, which has formed by growing over the yolk, is attached to the embryonic body by a narrow stalk. The yolk sac is highly vascular, its many blood vessels communicating with the blood channels of the embryo proper. The blood circulating through the vessels of the yolk sac carries dissolved yolk materials to all parts of the embryo, thus making the yolk available for chemical activities and growth in all regions.

From the hind region of the embryonic body, a sac bulges out on the underside and pushes its way between the yolk sac and chorion. This sac is the allantois (fig. 13.4). An appreciable part of the blood of the embryo is diverted into the allantois, where a rich system of blood vessels lies close to the inner surface of the shell. The porous shell permits the ready exchange of respiratory gases between the external air and the internal blood. The allantoic sac serves also as a receptacle for urinary wastes. Waste fluids excreted by the embryonic kidneys pass into the cavity of the allantois. The allantois thus has both respiratory and excretory functions. Finally, there is an outermost investing membrane, the chorion, which is abundantly supplied with blood vessels. The embryo depends, in part, on this vascularized enveloping membrane for carrying on gaseous exchange with the outer air through the porous shell.

The amnion, chorion, allantois, and yolk sac in the human embryo are essentially similar to these extraembryonic membranes in the reptiles (and birds). Curiously, the yolk sac does develop in the human embryo in spite of the absence of any appreciable amount of yolk. The human yolk sac, however, remains small and functionless. There is also no elaborate development of the allantois in the human embryo; the allantoic sac never becomes more than a rudimentary tube of minute size. Nevertheless, the very appearance of the yolk sac and allantois in the human embryo is one of the strongest pieces of evidence documenting the evolutionary relationships among the widely different kinds of vertebrates. To the student of evolution this means that the mammals, including man, are descended from animals that reproduced by means of externally laid eggs rich in yolk.

Adaptive Radiation of Reptiles

With the perfection of the amniotic egg, the reptiles exploited the wide expanses of land areas. The ancestral reptilian stock initiated one of the most spectacular adaptive radiations in life's history. The reptiles endured as dominant land animals of the earth for well over 100 million years. The

Mesozoic era, during which the reptiles thrived, is often referred to as the "Age of Reptiles."

Figure 13.5 reveals the variety of reptiles that blossomed from the basal

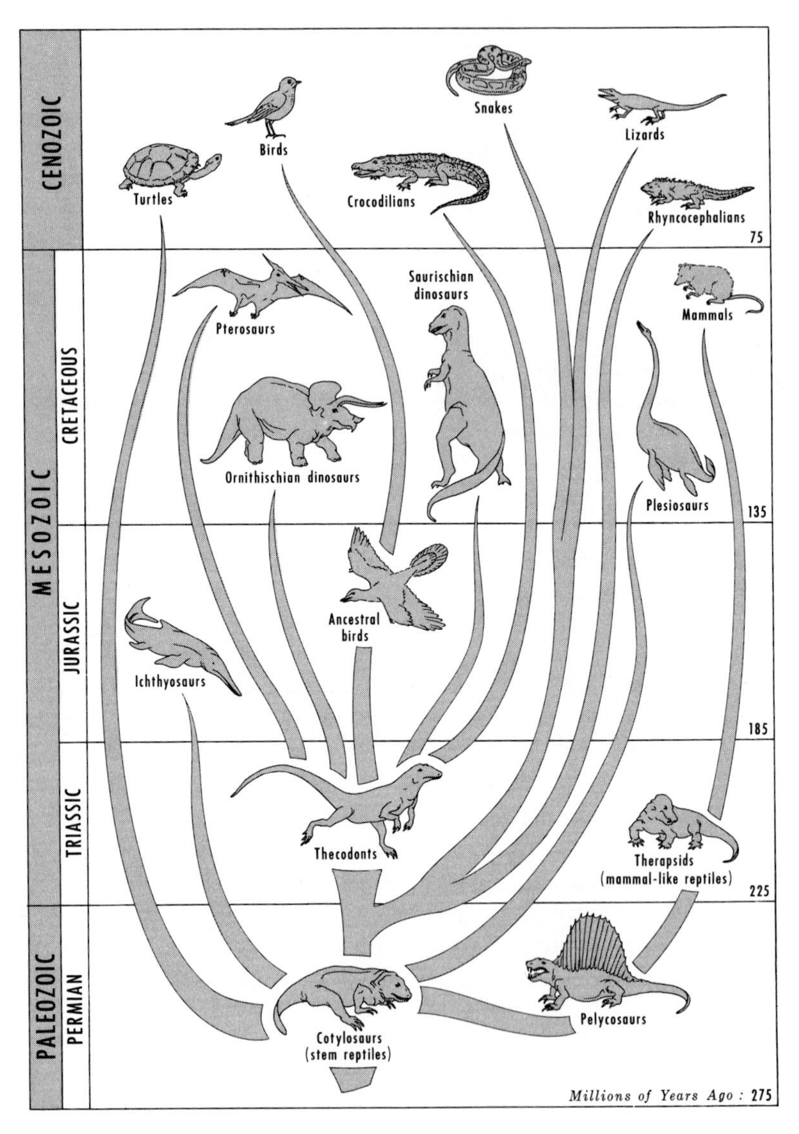

Figure 13.5 Adaptive radiation of reptiles. A vast horde of reptiles came into existence from the basal stock, the cotylosaurs, at the beginning of the Mesozoic era, roughly 225 million years ago. This matchless assemblage of reptiles was triumphant for a duration well over 100 million years. Then, before the close of the Mesozoic era, the great majority of reptiles passed into oblivion.

stock, the cotylosaurs. The dinosaurs were by far the most awe-inspiring and famous. They reigned over the lands until the close of the Mesozoic era before suffering extinction. The dinosaurs were remarkably diverse; they varied in size, bodily form, and habits. Some of the dinosaurs were carnivorous, such as the huge *Tyrannosaurus*, whereas others were vegetarians, such as the feeble-toothed but ponderous *Brontosaurus*. Not all dinosaurs were immense; some were no bigger than chickens. The dinosaurs were descended from the thecodonts—slender, fast-running, lizardlike creatures. In fact, there were two great groups of dinosaurs, the Saurischia and the Ornithischia (fig. 13.5), which evolved independently from two different lines of the thecodonts.

The thecodonts gave rise also to bizarre reptiles that took to the air, the pterosaurs. These "dragons of the air" possessed highly expansive wings and disproportionately short bodies. It seems as though the huge membranous wings developed at the sacrificial expense of the body. The winged pterosaurs succumbed before the end of the Mesozoic era. Another independent branch of the thecodonts led to eminently more successful flyers, the birds. The origin of birds from reptiles is revealed by the celebrated *Archaeopteryx*, a Jurassic form (fig. 13.5) that was essentially an air-borne lizard. This feathered creature possessed a slender, lizardlike tail and a scaly head equipped with reptilian teeth.

Certain reptiles returned to water. The streamlined, dolphinlike ichthyosaurs and the long-necked, short-bodied plesiosaurs were marine, fish-eating reptiles. These aquatic reptiles breathed by means of lungs; they did not redevelop the gills of their very distant fish ancestors. Among the early reptiles present at the beginning of Mesozoic days were the pelycosaurs, notable for their peculiar saillike extensions of the back (fig. 13.6). The function of the gaudy sail is unknown, but it should not be thought that this structural feature was merely ornamental or useless. As we have emphasized, traits of organisms were selected for their adaptive utility. Means of maintaining a stable body temperature are poorly developed in reptiles. It is not inconceivable that the pelycosauran sail was a functional device to achieve some degree of heat regulation. Be that as it may, the pelycosaurs gave rise to an important group of reptiles, the therapsids. These mammallike forms bridged the structural gap between the reptiles and the mammals (fig. 13.5).

Extinction and Replacement

The history of the reptiles attests to the ultimate fate of many groups of organisms—*extinction*. The reptilian dynasty collapsed before the close of the Mesozoic era. Of the vast host of Mesozoic reptiles, relatively few have survived to modern times; the ones that have include the lizards, snakes, crocodiles, and turtles. The famed land dinosaurs, the great marine plesiosaurs and ichthyosaurs, and the flying pterosaurs all became extinct. The

Figure 13.6 Dimetrodon, one of the pelycosaurs that flourished during Permian times. The gaudy "sail" may have served as a heat-regulating device. (From a restoration by Charles R. Knight; courtesy of the American Museum of Natural History.)

cause of the decline and death of the tremendous array of reptiles is obscure. The demise of the giant reptiles is generally attributed to their inability to adapt to some radical change in environmental conditions. In general, most species of organisms do become highly specialized. Perhaps many species pass unavoidably into oblivion when they cannot genetically adapt to radical environmental changes.

Whatever may be the cause of mass extinctions, the fact remains that as one group of organisms recedes and dies out completely, another group spreads and evolves. The decline of the reptiles provided evolutionary opportunities for the birds and the mammals. The vacancies in the habitats could then be occupied by these warm-blooded vertebrates. Small and inconspicuous during the Mesozoic era, the mammals arose to unquestionable dominance during the Cenozoic era, which began approximately 75 million years ago. The mammals diversified into marine forms (for example, the whale, dolphin, seal, and walrus), fossorial forms living underground (for example, the mole), flying and gliding animals (for example, the bat and flying squirrel), and cursorial types well-adapted for running (for example, the horse).

The various mammalian groups are adapted to their different modes of life. The appendages, in particular, are specialized for flight, swimming, or movement on land. An important lesson may be drawn from the variety of specialized appendages. Superficially there is scant resemblance between the arm of a man, the flipper of a whale, and the wing of a bat. And yet, a close comparison of the skeletal elements (fig. 13.7) shows that the structural design, bone for bone, is basically the same. The differences are mainly in the relative lengths of the component bones. In the forelimb of the bat, for instance, the metacarpals and phalanges (except those of the thumb) are greatly elongated. Although highly modified, the bones of the bat's wing are not fundamentally different from those of other mammals. The conclusion is inescapable that the limb bones of man, the bat, and the whale are modifications of a common ancestral pattern. The facts admit no other logical interpretation. Indeed, as seen in figure 13.7, the forelimbs of all tetrapod vertebrates exhibit a unity of anatomical pattern intelligible only on the basis of common inheritance. The corresponding limb bones of tetrapod vertebrates are said to be *homologous,* since they are structurally identical with those in the common ancestor. In contrast, the wing of a bird and the wing of a butterfly are *analogous*; both are used for flight, but they are built on an entirely different structural plan.

Evolution of Horses

The cardinal feature of adaptive radiation is the emergence from a central, generalized stock of a large number of divergent branches or lineages. Not all branches persist; indeed, the general rule is that all but a few perish. The disappearance of many branches in the distant past may lead the observer today to the mistaken impression that the evolution of a particular group was not at all intricately forked. Thus, the evolution of horses is commonly, but erroneously, depicted as an undeviating, straight-line progression from the small, terrier-sized *Hyracotherium* (formerly known as *Eohippus,* the "dawn horse") to the large modern horse, *Equus*. On the contrary, the detailed work of modern paleontologists, prominent among them George Gaylord Simpson, has revealed convincingly a pattern of many divergent lineages. The major lineages directly or indirectly involved in the emergence of the modern horse are shown in figure 13.8.

The ancestry of horses can be traced back to the beginning of the Eocene epoch, some 60 million years ago. The diminutive *Hyracotherium* was about 10 inches high and had four functional toes on the front foot and three on the hind. It was not adapted for speed and browsed on the soft leaves and fruits of bushes. This dawn horse could scarcely have grazed

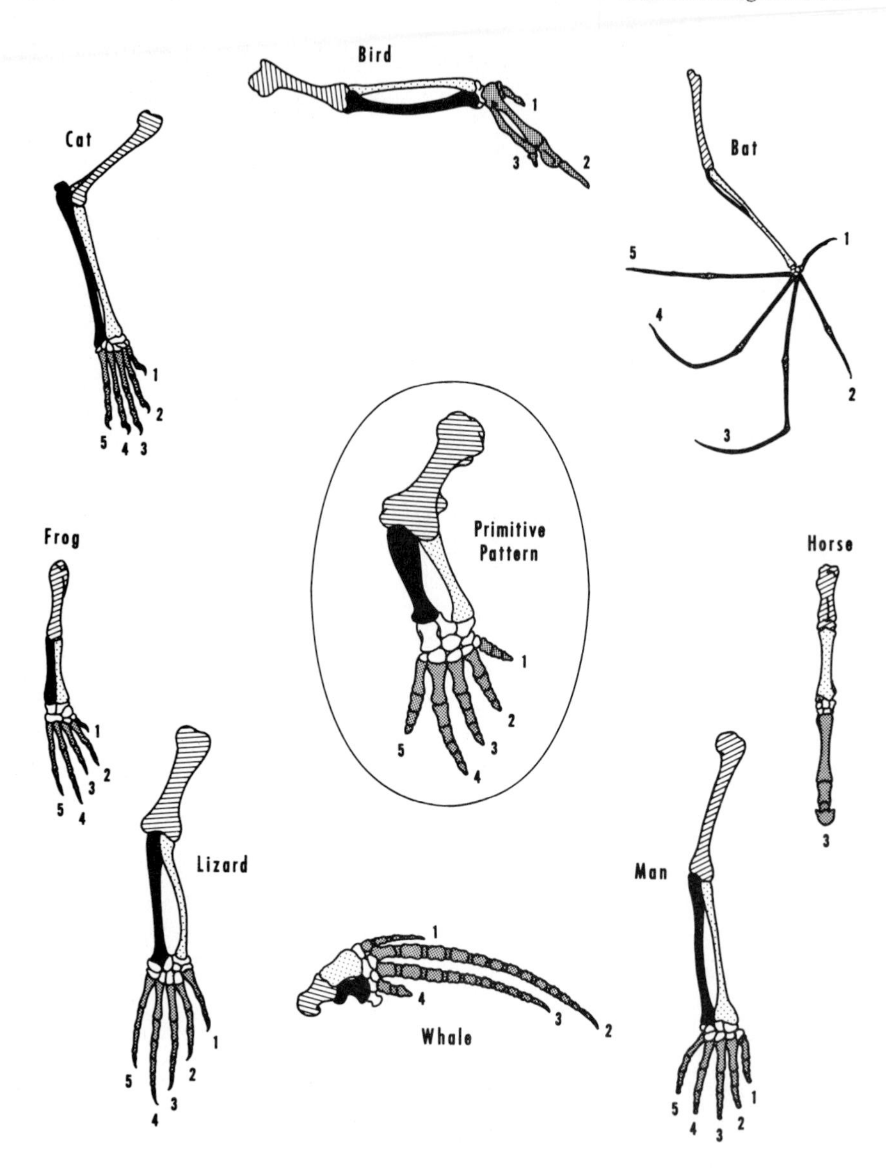

Figure 13.7 Varied forelimbs of vertebrates, all of which are built on the same structural plan. The best explanation for the fundamentally similar framework of bones is that man and all other vertebrates share a common ancestry. Homologous bones are indicated as follows: humerus (upper arm)—crosshatching; radius (forearm)—light stippling; ulna (forearm)—black; carpals (wrist)—white; metacarpals (palm) and phalanges (digits)—heavy stippling. The number of phalanges in each digit is indicated by a numeral, beginning with the first digit (thumb).

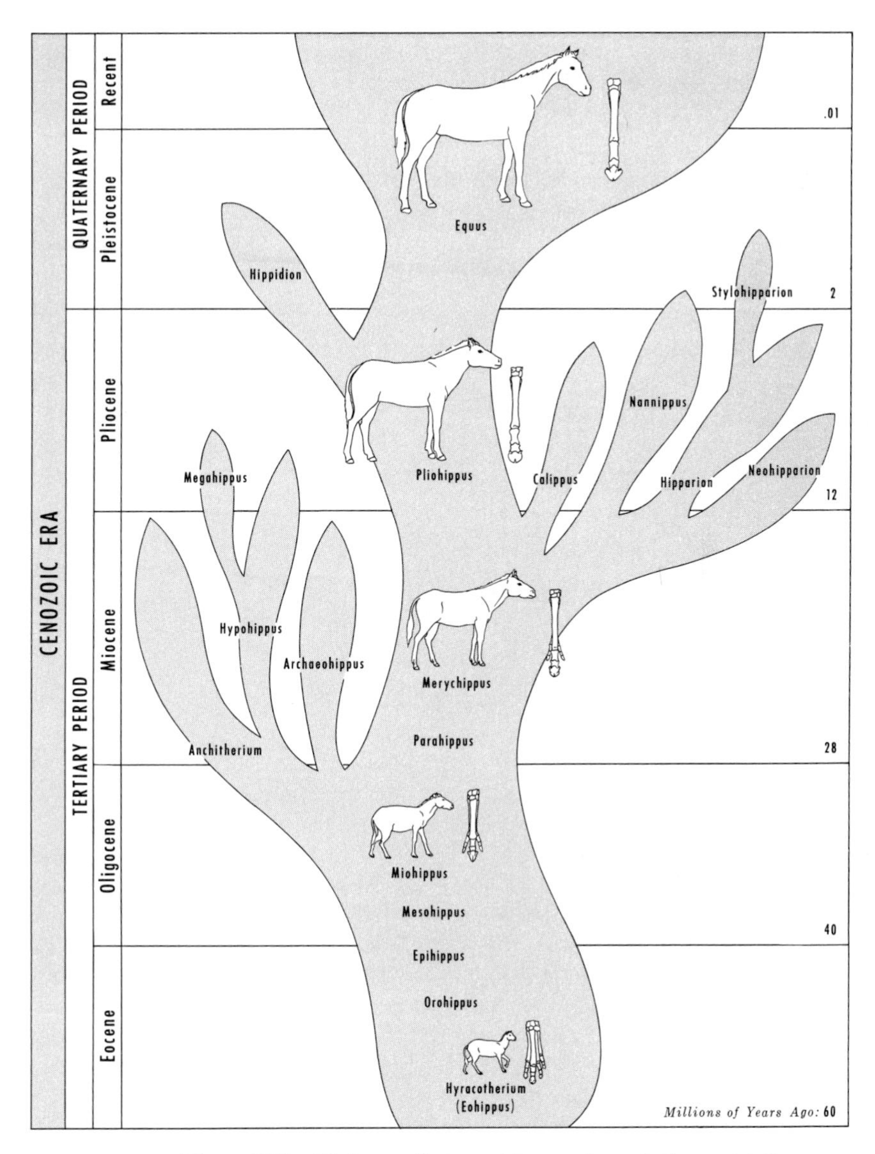

Figure 13.8 Phylogenetic tree of horses through time, with its many divergent branches. All branches died out, save one which eventually culminated in the modern group of horses, *Equus*. The history of horses dates back to early Cenozoic times, some 60 million years ago. (The Cenozoic *era* is divided into *periods,* which in turn are subdivided into *epochs.*)

on harsh grass, as its molar teeth had low crowns and very weak grinding surfaces. *Hyracotherium* spread over North America and Europe, but the European assemblage left no descendants. The derivatives of the North American dawn horse in the Eocene were *Orohippus* and *Epihippus,* and these gradually gave way to more progressive kinds of horses in the Oligocene, *Mesohippus* and *Miohippus*. Both of these Oligocene horses were about 24 inches high with only three toes on each foot, but they remained browsers with low-crowned teeth. The *Miohippus* stock subsequently radiated into a variety of types that lived in the Miocene epoch. Most of the Miocene offshoots (for example, *Anchitherium* and *Hypohippus*) continued the browsing habits of their ancestors. However, one stock—*Merychippus*—had successfully exploited a new mode of life associated with the expansion of grasslands in Miocene times. *Merychippus* was the first true grazing form, well equipped with high-crowned, complexly ridged teeth effective in grinding stout grass.

From the slow-footed, three-toed *Merychippus* emerged the fleet-footed, one-toed *Pliohippus* during the Pliocene epoch. The appearance of the one-toed condition was a landmark in horse history, an event doubtlessly fostered by natural selection. Grazing horses in the open, grassy plains were subject to appreciable predation from carnivores. Natural selection favored reduction of the toes, lightening the legs for speed. However, not all descendants of *Merychippus* evolved the progressive single-toed condition. Several lines of Pliocene horses, such as *Hipparion* and *Nannippus,* retained the conservative three-toed pattern. These conservative Pliocene horses represent evolutionary blind alleys, removed from the main line of ancestry of the modern horse (fig. 13.8). Only *Pliohippus* the first one-toed horse, remained to become the progenitor of the modern form, *Equus*. The modern horse arose in North America at the close of the Pliocene epoch, and spread rapidly over most of the world. *Equus* inexplicably vanished in North America a few thousand years ago, before the arrival of the first white colonist. The modern horse was reintroduced from Europe in the early 1500s by the Spaniards. All present-day horses are domestic; it is doubtful whether there are any existing stocks of truly wild horses that are not descendants of once-domesticated populations.

Convergence

We have learned that a population of organisms tends to become highly diversified when it spreads over an area with varied habitats. The population can diverge into radically different lines, each modified for a specific ecological role. Now, essentially similar habitats may be found in widely separate parts of the world. It would thus not be surprising to find that two groups of organisms, unrelated by descent but living under similar environmental conditions in different geographic regions, can exhibit similarities in habits and general appearances. The tendency of one group of organisms to develop superficial resemblances to another group of different ancestry is called *convergence.*

Convergence is not an uncommon phenomenon in nature. Many unrelated, or remotely related, organisms have converged in appearance as a consequence of exploitation of habitats of similar ecological makeup. The relationship of convergence to adaptive radiation should be evident; the former is the inevitable result of the countless series of adaptive radiations that have taken place in scattered parts of the globe. A striking example of convergence is afforded by the living marsupials of Australia.

The massive island continent of Australia has long been isolated from Asia, at least since the commencement of the Cenozoic era, the last 75 million years. It is on this isolated island that the marsupials—primitive pouched mammals—survived, free from the competition of the more efficient placental mammals when they came into prominence. In the absence of a land migration route between Asia and Australia, the latter land mass was inaccessible to practically all the placental mammals. On other continents, the marsupials perished, save for the peculiar American opossums, the didelphids.

Imprisoned in Australia, the marsupials spread into a variety of habitats. Several live in the open plains and grasslands; some are tree-dwellers; others are burrowers; and still others are gliders (fig. 13.9). Most kangaroos are terrestrial, but one variety, the monkeylike kangaroo, is arboreal. The slow-moving, nocturnal "teddy bear," or koala, lives and feeds on *Eucalyptus,* the dominant tree of Australia. The bandicott, with rabbitlike ears, has sturdy claws adapted for digging in the ground in search of insects. Marsupial moles live in desert burrows, and flying phalangers have webs of loose skin stretched between the forelimbs and hindlimbs. The flying phalanger cannot actually fly, but is adept at gliding. The impressive diversity of marsupials thus represents an admirable example of adaptive radiation.

The marsupials of Australia also vividly illustrate the phenomenon of convergence. They have filled the ecological niches normally occupied by placental mammals in other parts of the world. The marsupial "mouse," "mole," "anteater," "wolf," flying phalanger, and groundhoglike wombat strikingly resemble the true placental types—mouse, mole, anteater, wolf, flying squirrel, and groundhog, respectively—in general appearance and in ways of life.

It is interesting to note that a marsupial "bat" has not evolved in Australia. The opportunity was apparently denied by the invasion of placental bats from Asia, one of the few placental forms that managed, probably as a result of dispersal by flight and winds, to reach Australia. With the coming of man, the secure existence of the marsupials has been threatened. Prehistoric man introduced dogs, which ran wild (the dingos); later human settlers brought a number of European placental mammals, such as the rabbit, hare, fox, and Norway rat. Among the marsupials faced with extinc-

Figure 13.9 Marsupials, or pouched mammals, of Australia, illustrating the twin themes of adaptive radiation and convergence. The marsupials have radiated, or diversified, into a variety of forms, ranging from the tiny insect-eating jumping "mouse" to the fierce, flesh-eating Tasmanian "wolf." Many of the marsupials resemble in appearance and habit the placental mammals of other parts of the world, although they are *not* closely related to them.

tion are the marsupial "wolf," which survives today only in Tasmania and the slow-moving "anteater," which is rapidly disappearing.

Selected Readings

Colbert, E. H. 1955. *Evolution of the vertebrates.* New York: John C. Wiley & Sons.

Colbert, E. H. 1961. *Dinosaurs, their discovery and their world.* New York: E. P. Dutton & Co.

Dobzhansky, T. 1955. *Evolution, genetics, and man.* New York: John C. Wiley & Sons.

Hotton, N. 1968. *The evidence of evolution.* New York: American Heritage Publishing Co.

Johansen, K. 1968. Air-breathing fishes. *Scientific American,* October, pp. 102–11.

Kurtén, B. 1969. Continental drift and evolution. *Scientific American,* March, pp. 54–64.

Newell, N. D. 1963. Crises in the history of life. *Scientific American,* February, pp. 76–92.

Romer, A. S. 1959. *The vertebrate story.* Chicago: University of Chicago Press.

Simpson, G. G. 1951. *Horses.* New York: Oxford University Press.

Simpson, G. G. 1953. *Life of the past.* New Haven, Conn.: Yale University Press.

Smith, H. W. 1961. *From fish to philosopher.* Garden City, N.Y.: Doubleday & Co.

Stirton, R. A. 1959. *Time, life, and man.* New York: John C. Wiley & Sons.

Young, J. Z. 1950. *The life of vertebrates.* New York: Oxford University Press.

14

Origin and History of Life

We have been unceasingly taught not to believe in spontaneous generation, the view that living things can originate from lifeless matter. The classical experiments of Francesco Redi in 1688, Lazzaro Spallanzani in 1765, and Louis Pasteur in 1862 provided proof that new life can only come from existing life. However, these experiments revealed only that life cannot arise spontaneously under conditions that exist on earth today. Conditions on the primeval earth billions of years ago were assuredly different from present conditions, and the first form of life, or self-duplicating particle, did arise spontaneously from chemical inanimate substances. It should be clear that the conditions under which life originated were unique; the reorigin of life on present-day earth is scarcely possible.

Primitive Earth

The view that life emerged by a long and gradual process of chemical evolution was first convincingly set forth by the Russian biochemist Alexander I. Oparin, in 1924, in an enthralling book entitled *The Origin of Life*. The transformation of lifeless chemicals into living matter extended over some 2 billion years. Such a transformation, as Oparin points out, is no longer possible today. If by pure chance a living particle approaching that of the first form of life should now appear, it would be rapidly decomposed by the oxygen of the air or quickly destroyed by the countless microorganisms presently populating the earth.

Our earth is estimated to be 5 billion years old. It was formerly thought that the earth originated as a fiery mass which was torn away from the sun. Astronomers now generally acknowledge that the earth (like other planets in the solar system) condensed out of a whirling cloud of gas surrounding the primitive sun. The atmosphere of the pristine earth was quite unlike our present atmosphere. Oxygen in the free gaseous state was virtually absent; it was bound in water and metallic oxides. Accordingly, any complex organic compound that arose during this early time would not be subject to degradation by free oxygen.

The early gas cloud was especially rich in hydrogen. The hydrogen (H_2) of the primordial earth chemically united with carbon to form methane (CH_4), with nitrogen to form ammonia (NH_3), and with oxygen to form water vapor (H_2O). Thus, the early atmosphere was a strongly reducing (nonoxygenic) one, containing primarily hydrogen, methane, ammonia, and water. The atmospheric water vapor condensed into drops and fell as rain; the rains eroded the rocks and washed minerals (such as chlorides and phosphates) into the seas. The stage was set for the combination of the varied chemical elements. Chemicals from the atmosphere mixed and reacted with those in the waters to form a wealth of hydrocarbons (that is, compounds of hydrogen and carbon). Water, hydrocarbons, and ammonia are the raw materials of amino acids, which, in turn, are the building blocks for the larger protein molecules. Thus, in the primitive sea, amino acids accumulated in considerable quantities and were joined together to form proteins.

Complex carbon compounds such as amino acids and proteins are termed *organic* because they are made by living organisms. Our present-day green plants use the energy of sunlight to synthesize organic compounds from simple molecules. What, then, was the energy source in the primitive earth, and how was synthesis of organic compounds effected in the absence of living things? It is generally held that ultraviolet rays from the sun, electrical discharges such as lightning, and dry heat from volcanic activity furnished the means to join the simple carbon compounds and nitrogenous substances into amino acids. Is there a valid basis for such a widely accepted view?

Experimental Synthesis of Organic Compounds

In the early days of chemistry, it was believed that organic compounds could be produced only by living organisms. But in 1828, Friedrich Wöhler succeeded in manufacturing the organic compound *urea* under artificial conditions in the laboratory. Since Wöhler's discovery, a large variety of organic chemicals (amino acids, monosaccharides, purines, and vitamins) formerly produced only in organisms have been artificially synthesized.

In 1953, Stanley Miller, then at the University of Chicago and a student of Nobel laureate Harold Urey, synthesized organic compounds under conditions resembling the primitive atmosphere of the earth. He passed electrical sparks through a mixture of hydrogen, water, ammonia, and methane. The electrical discharges duplicated the effects of violent electrical storms in the primitive universe. In the laboratory, the four simple inorganic molecules interacted, after a mere week, to form several kinds of amino acids, among them alanine, glycine, aspartic acid, and glutamic acid. Miller's instructive experiment has been successfully repeated by a number of investigators; amino acids can also be obtained by irradiating a similar mixture of gases with ultraviolet light.

The synthesis of amino acids is only a small step toward the synthesis of a living cell. In 1964, Sidney W. Fox, then at Florida State University and later at the University of Miami, reasoned that proteins were synthesized from amino acids in the primitive earth by thermal energy, or heat. He accordingly heated a mixture of 18 amino acids to temperatures of 160–200°C for varying periods of time. He obtained stable, protein-like macromolecules, which he termed *proteinoids.* These thermally produced proteinoids are similar to natural proteins in many respects. As a striking instance, bacteria can actually utilize the proteinoids in a culture medium, degrading them enzymatically into individual amino acids. Equally important, when the proteinoid material was cooled and examined under a microscope, Fox observed small, spherical, cell-like units that had arisen from aggregations of the proteinoids. These molecular aggregates, called *microspheres,* exhibited a general resemblance to spherical bacteria. Such microspherelike aggregates could have been the forerunners of the first living organism. However, there remains a large hiatus between the formation of organic microspheres and the appearance of the first living cell, capable of specific catalytic functions, reproduction, and mutation.

Life's Beginnings

It is evident that organic compounds can be formed without the intervention of living organisms. Thus, it appears likely that the sea of the primitive earth spontaneously accumulated a great variety of organic molecules. The sea became a sort of dilute organic soup (an aquatic Garden of Eden), in which the molecules collided and reacted to form new molecules of increasing levels of complexity. Proteins capable of catalysis, or enzymatic activity, had to evolve, and nucleic acid molecules capable of self-replication must also have developed. However, the incisive question has yet to be settled—whether the initial process involved first the appearance of the machinery for self-replication (that is, self-duplicating nucleic acids) followed by the development of the cytoplasm and membrane, or whether a cytoplasm with internal organization and metabolic capacities preceded the nuclear mechanism. Those who advocate a naked gene (DNA) as the first living unit must reconcile their view with the fact that a gene, like a virus, requires a well-coordinated metabolic system to function. We may recall that a virus is dependent on living cells for its existence; hence, parasitic viruses could not have arisen before cells had evolved.

We may assume that the first living systems drew upon the wealth of organic materials in the sea broth. Organisms that are nutritionally dependent on their environment for ready-made organic substances are called *heterotrophs* (Greek, *hetero,* "other" and *trophos,* "one that feeds"). The primitive one-celled heterotroph probably had little more than a few genes, a few proteins, and a selectively permeable cell membrane. The heterotrophs multiplied rapidly in an environment with a copious supply of dissolved organic substances. However, the ancient heterotrophs could survive only as long as the existing store of organic molecules lasted. Eventually, living systems had to evolve the ability to synthesize their own organic requirements from simple inorganic substances. In the course of time, *autotrophs* (*auto,* meaning "self") arose, which could manufacture organic nutrients from simpler molecules.

The first simple autotroph arose in an anaerobic world, one in which little, if any, free oxygen was available. The primitive autotrophs obtained their energy from the relatively inefficient process of fermentation (the breakdown of organic compounds in the absence of oxygen). Thus, the early fermentative autotrophs were much like our present-day anaerobic bacteria and yeast. The metabolic processes of the anaerobic autotrophs resulted in the liberation of large amounts of carbon dioxide into the atmosphere. Once this occurred, the way was paved for the evolution of organisms that could use carbon dioxide as the sole source of carbon in synthesizing organic compounds and could use sunlight as the sole source of energy. Such organisms would be the photosynthetic cells.

Early photosynthetic cells probably split hydrogen-containing compounds such as hydrogen sulfide. In other words, as is still performed today by sulfur bacteria, hydrogen sulfide is cleaved into hydrogen and sulfur. The hydrogen is used by the cell to synthesize organic compounds, and the sulfur is released as a waste product (as evidenced by the earth's great sulfur deposits). In time, the process of photosynthesis was refined so that water served as the source of hydrogen. The result was the release of oxygen as a waste product. At this stage, free oxygen became established for the first time in the atmosphere.

The first organisms to use water as the hydrogen source in photosynthesis were the blue-green algae. Since blue-green algae were active photosynthesizers, atmospheric oxygen accumulated in increasing amounts. Many primitive anaerobic bacteria, incapable of adapting to free oxygen, remained in portions of the environment that were anaerobic, such as sulfur springs and oxygen-free muds. New kinds of bacteria, however, arose that were capable of utilizing the free oxygen. Today, there are bacterial types that are anaerobic as well as aerobic.

The earth's atmosphere gradually changed from a reducing, or hydrogen-rich, atmosphere to an oxidizing, or oxygen-rich, atmosphere. The rising levels of atmospheric oxygen set the stage for the appearance of one-celled eucaryotic organisms, between 600 million and 1 billion years ago (fig. 14.1). Then, within the comparatively short span of 600 million years, the one-celled eucaryotes evolved in various directions to give rise to a wealth of multicellular life forms inhabiting the earth.

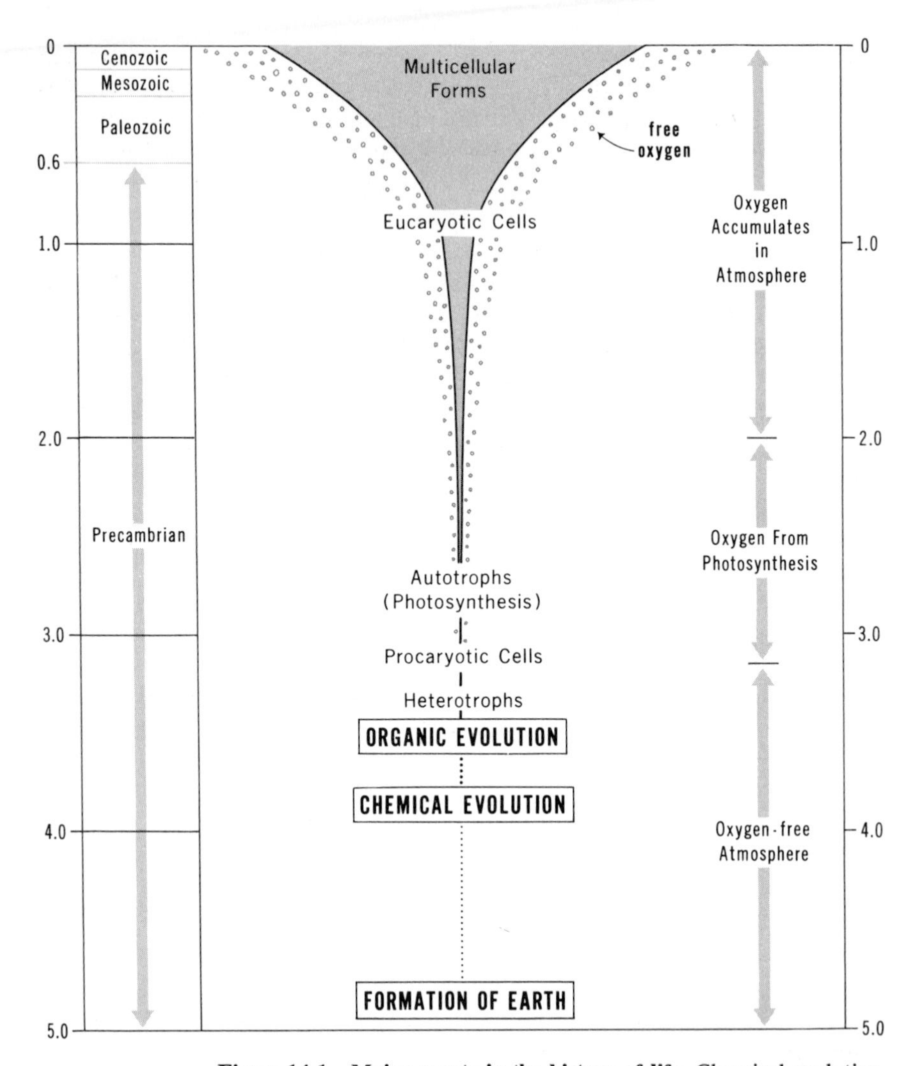

Figure 14.1 Major events in the history of life. Chemical evolution preceded biological evolution. The first self-duplicating life forms were heterotrophs. Oxygen began to accumulate in the atmosphere with the appearance of photosynthetic autotrophs. Rising levels of atmospheric oxygen were associated with the emergence of eucaryotes. The last 600 million years has witnessed a great diversity of life in an oxygen-rich atmosphere.

Evolution of Eucaryotes

The simplest cells in nature—the bacteria and blue-green algae—have been classified as *procaryotic* cells. These cells have no nuclear membrane by which the hereditary materials (DNA) are set apart from the cytoplasm, and lack specialized cytoplasmic bodies (organelles) such as mitochondria and chloroplasts. In a bacterial cell, the DNA forms a simple closed loop that is attached to the inside of the cell's membrane. In contrast, the cells of higher plants and animals, or *eucaryotic* cells, have a distinct nuclear membrane (that encloses strands of DNA) as well as an elaborate system of membrane-bound cytoplasmic organelles. In the eucaryotic plant cell, a prominent organelle is the chloroplast, the photosynthesizing particle that contains light-absorbing pigment. In eucaryotic cells of both plants and animals, mitochondria are organelles concerned with energy production.

Mitochondria have a number of interesting properties that suggest that they were once free-living, or independent, bacterialike organisms. Mitochondria have small amounts of their own DNA; this DNA exists as a loop-shaped molecule like the DNA of bacteria. Mitochondria also possesses the genetic capacity to incorporate amino acids into proteins. These properties that the mitochondria exhibit have led to the hypothesis that mitochondria may have been derived from primitive aerobic bacteria that were engulfed by predatory organisms, probably fermentative bacteria, destined to become eucaryotic (fig. 14.2). The predatory hosts became dependent on their enslaved mitochondria, and the latter, in turn, became dependent on their hosts. Thus, the association was of mutual advantage to the predatory cell and the engulfed prey. Such a close association, or partnership, is called *symbiosis*. In essence, the mitochondria (formerly oxygen-respiring bacteria) established permanent residence within the hosts, which became the first eucaryotic cells.

Eucaryotic plant cells contain both mitochondria and chloroplasts. Just as aerobic bacterial cells can be equated with mitochondria, blue-green algae cells are basically equivalent to the chlorophyll-containing chloroplasts. Chloroplasts, like mitochondria, have their own unique DNA and the associated protein-synthesizing machinery. We can now speculate that those predatory cells that engulfed but did not digest the aerobic bacterial cells became the eucaryotic animal cells. Those predators that captured both the photosynthetic blue-green algal cells and aerobic bacteria evolved into eucaryotic plant cells (fig. 14.2). Accordingly, the engulfed prey became permanent symbiotic residents—either as mitochondria or chloroplasts—within the predatory cell. The idea that present-day mitochondria and chloroplasts may be descendants of ancestral aerobic bacteria and blue-green algae, respectively, is highly speculative but provocative.

Some authors regard the relatively abrupt origin of the organelles of eucaryotes by a process of symbiosis as improbable. The evolution of living organisms normally proceeds through a series of gradual changes. Continual modification of the surface membrane may have been the basic evolutionary mechanism in the differentiation of eucaryotic cell from procaryotic cell. The organelles of eucaryotic cells may have evolved by the invagination, or drawing inward, of the surface membrane of a primitive cell. Figure 14.3

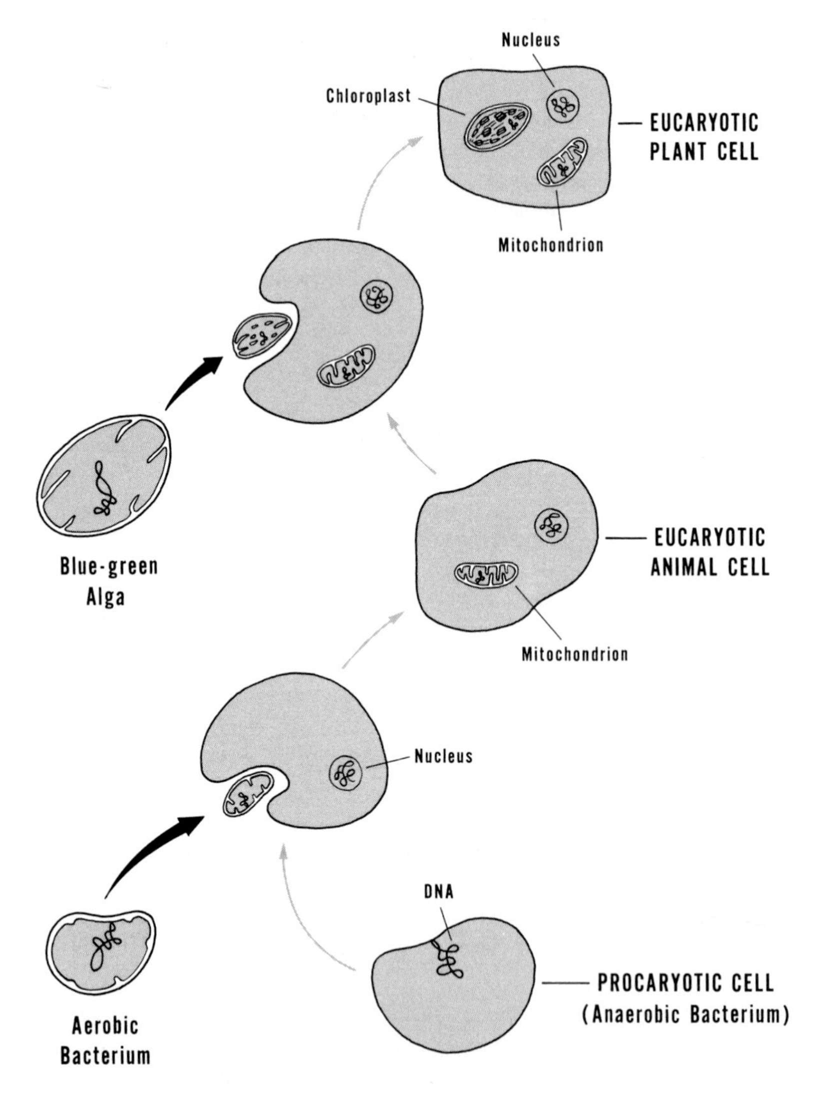

Figure 14.2 Evolution of eucaryotes through symbiosis. The eucaryotic animal cell may have arisen through a symbiotic relationship between a procaryotic cell and an aerobic bacterium. The eucaryotic plant cell may have originated by a comparable symbiotic relationship between a blue-green alga and a eucaryotic animal cell.

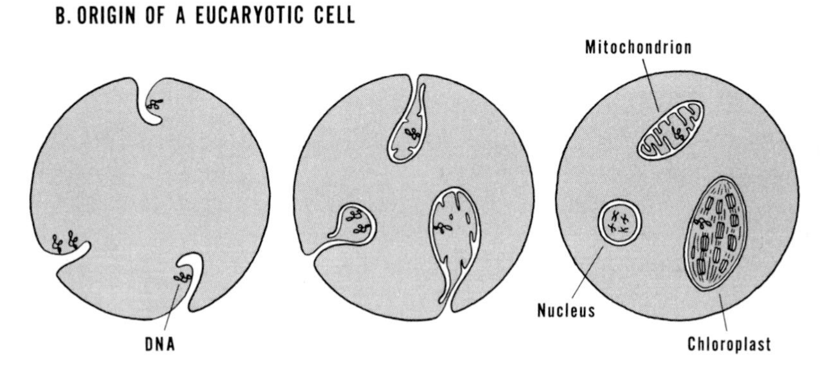

Figure 14.3 Hypothetical scheme of the development of the mitochondrion from the surface membrane (*A*), and of the origin of a eucaryotic cell by the invagination, or drawing inward, of the surface membrane in several places (*B*).

shows the postulated mechanism for the origin of a mitochondrion from the cell surface membrane. In support of this scheme, it is interesting that the enzymes for the breakdown of carbohydrate for energy production in bacteria are incorporated in the structure of the cell surface membrane. In like manner, other specialized internal structures of eucaryotes, including the membrane-enclosed nucleus, may have differentiated from invaginated cell surfaces (fig. 14.3). This hypothesis is simpler than the origin of eucaryotic organelles by symbiosis, and is considered more plausible by several investigators.

Fossils and the Historical Record

An organism becomes preserved as a fossil when it is trapped in soft sediment that settles at the bottom of a lake or ocean. The deposits of mud and sand harden into the sedimentary, or stratified, rocks of the earth's crust. Fossils, then, are the remains of past organisms preserved or imprinted in sedimentary rocks. Not all fossils are mere impressions of buried parts. In the case of petrified wood, for example, the wood had become infiltrated with mineral substances that crystallized and hardened. Some components of organisms are resistant to decay, such as the silica walls of diatoms, which accumulate to form diatomaceous earth.

The sample of past life is incomplete and uneven. In many places of the earth, the sedimentary rocks have been so subjected to pressure and heat after their deposition that the fossils in them have been destroyed. Moreover, organisms with soft tissues (jellyfishes and insects, for example) are not favorable for preservation. In contrast, the woody parts of plants, the shells of mollusks, and the bones of vertebrates are relatively common as fossils. Despite the imperfections of the fossil record, the available fossil assemblage contains an extraordinary amount of information. The older strata of rock, those deposited first, bear only relatively simple kinds of life, whereas the newer or younger beds contain progressively more and more complex types of life. Each species of organism now living on the earth has developed from an earlier and simpler ancestral form.

Geologists and paleontologists have divided the earth's past history into five main divisions, or *eras,* associated with five major rock strata. The eras embrace a number of subdivisions, or *periods,* and the periods are further subdivided into *epochs*. The most ancient era is the Archeozoic, followed by the Proterozoic, Paleozoic, Mesozoic, and lastly, the Cenozoic, the era of recent types of life. Fossils appear in abundance at the beginning of the Paleozoic era, technically the Cambrian period, 600 million years ago. However, as noted earlier, at least 3 billion years of slow organic evolution preceded the diversity of life of the Cambrian period. In recent years, explorations of Precambrian rocks have uncovered the remains of procaryotic bacteria and blue-green algae.

One of the most remarkable discoveries of Precambrian life was reported in 1965 by the micropaleontologists Elso S. Barghoorn and J. William Schopf of Harvard University. Electron microscopic examinations of samples of deep sedimentary rock from South Africa (the so-called Fig Tree sediments) revealed remnants of rod-shaped, bacterialike organisms that existed 3.2 billion years ago. In addition, minute traces of two chemical substances, phytane and pristane, have been found in the Fig Tree rocks. These chemicals are the relatively stable breakdown products of chlorophyll. Thus, photosynthesis by algae, in which oxygen is released, may have begun about 3 billion years ago. Other ancient rocks—such as the 2-billion-year-old Gunflint rock formation in Ontario, Canada—contain an assemblage of fossil microorganisms that appear to be blue-green algae, possibly red algae, and even some fungi. Eucaryote organization is clearly in evidence in deposits that are approximately 1 billion years old. The Bitter Springs formation of Australia contains eucaryote fossils that represent higher algae and fungi.

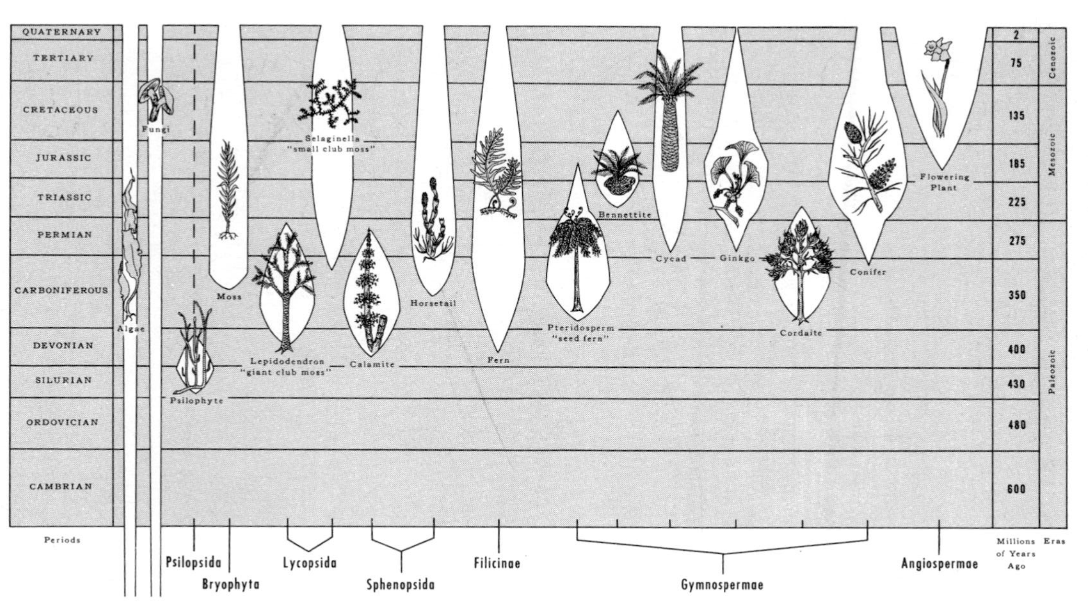

Figure 14.4 Historical record of plant life. Plant life during the Cambrian and Ordovician, the first two periods of the Paleozoic era, was confined to the water; seaweeds (algae) of immense size, often several hundred feet in length, dominated the seas. Land plants came into existence in Silurian time, in the form of strange little vascular plants, the psilophytes. In the Carboniferous period, imposing spore-bearing trees (Lepidodendrids and Calamites) and primitive naked-seeded plants (Pteriodosperms and Cordaites) reached their peak of development. The end of the Paleozoic era marked the extinction of the majority of the luxuriant trees of the Carboniferous coal swamps. The Mesozoic era was the *Age of Gymnosperms*, as evidenced by the abundance of cycads, ginkgoes, and conifers. Flowering plants (Angiosperms) rose to ascendancy toward the close of the Mesozoic era and established themselves as the dominant plant group of the earth.

The transition from procaryote to eucaryote organization apparently occurred during the late Precambrian.

The general character of plant life from the dawn of the Cambrian period to recent times is depicted in figure 14.4. Only the broader aspects of past evolution are portrayed. It is evident that the flora has changed in composition over geologic time. New types of plants have continually appeared. Some types thrived for a certain time and then disappeared. Others arose and continued to flourish. Still others have survived until the present only in much reduced numbers.

The Cambrian and Silurian periods were characterized by a diverse group of algae in the oceans and the seas. One of the most significant advances, which occurred during the Silurian period, was the transition from aquatic existence to life on land. The first plants that established themselves on land were diminutive herbaceous forms, the *psilophytes,* literally "naked plants," in allusion to their bare, leafless stems. Their existence made possible the subsequent emergence of an infinite variety of tree-sized plants that flourished in the swamps of the Carboniferous times. Carboniferous forests included the giant club-moss, *Lepidodendron,* the tall *Calamites,* the coarse-leaved *Cordaites,* and the "seed ferns," *Pteridosperms,* which were the first seed-bearing plants and not true ferns at all. These ancient groups dwindled toward the close of the Paleozoic era and shortly became extinct. Their decomposed remains led to the formation of extensive coal beds throughout the world. Figure 14.5 shows the appearance of a Carboniferous swamp as reconstructed from fossils.

Figure 14.5 Luxuriant forests of giant trees with dense undergrowth flourished in the Great Coal Age (Carboniferous period), between 350 million and 275 million years ago. The massive trunks at the left are the Lepidodendrids, an extinct group whose modern relatives include the small, undistinguished club mosses (*Selaginella*) and the ground pines (*Lycopodium*). The tall, slender tree with whorled leaves at the right is a Calamite, represented today by a less prominent descendant, the horsetail *Equisetum.* The fernlike plants bearing seeds (at the left) are seed ferns (Pteridosperms), which resembled ferns but were actually the first true seed-bearing plants (Gymnosperms). (Courtesy of the Chicago Natural History Museum.)

By far the greater number of Paleozoic species of plants failed to survive. The Paleozoic flora was largely replaced by the seed-forming gymnosperms, which became prominent in the early Mesozoic era. A diverse assemblage of cycads, ginkgoes (maidenhair trees), and conifers formed elaborate forests. During the closing years of the Mesozoic, the flowering plants (angiosperms), which began very modestly in the Jurassic period, underwent a phenomenal development and constitute today the dominant plants of the earth. The angiosperms have radiated into a variety of habitats, from sea level to mountain summits and from the humid tropics to the dry deserts. Associated with this diversity of habitat is great variety in general form and manner of growth. Many angiosperms have reverted to an aquatic existence. The familiar duckweed, which covers the surface of a pond, is a striking example.

In the animal kingdom, we witness a comparable picture of endless change (fig. 14.6). The deep Cambrian rocks contain the remains of varied marine invertebrate animals—sponges, jellyfishes, worms, shellfishes, starfishes, and crustaceans. These invertebrates were already so well developed that their differentiation must have taken place during the long period preceding the Cambrian. That this is actually the case has been revealed by the important finding, by the Australian geologists R. C. Sprigg and Martin F. Glaessner, of a rich deposit of Precambrian fossils in the Ediacara Hills in southern Australia. The fossil-bearing rocks lie well below the oldest Cambrian strata, and contain imprints of jellyfishes, tracks of worms, traces of soft corals, and other animals of uncertain nature. Additionally, the Dartmouth University Geologist Andrew McNair uncovered Precambrian remains of invertebrate animals (worms and brachiopods) in rocks, 700 million years old, on Victoria Island in the Canadian Arctic.

Dominating the scene in early Paleozoic waters were bizarre arthropods, the trilobites and the large scorpionlike eurypterids. Common in all Paleozoic periods were the nautiloids, related to the modern nautilus, and the brachiopods, or lampshells, relatively inconspicuous today. The odd graptolites, colonial animals whose carbonaceous remains resemble pencil marks, attained the peak of their development in the Ordovician period and then abruptly declined. No land animals are known for Cambrian and Ordovician times. Seascapes of the early Paleozoic are shown in figure 14.7.

Many of the prominent Paleozoic marine invertebrate groups became extinct or declined sharply in numbers before the Mesozoic era. During the Mesozoic, shelled ammonoids flourished in the seas, and insects and reptiles were the predominant land forms. At the close of the Mesozoic, the once-successful marine ammonoids perished and the reptilian dynasty collapsed, giving way to the birds and mammals. Insects have continued to thrive and have differentiated into a staggering variety of species. Well over 600,000 different species of insects have been described, and conservative estimates place the total number of living species today at 3 million.

During the course of evolution, plant and animal groups interacted to

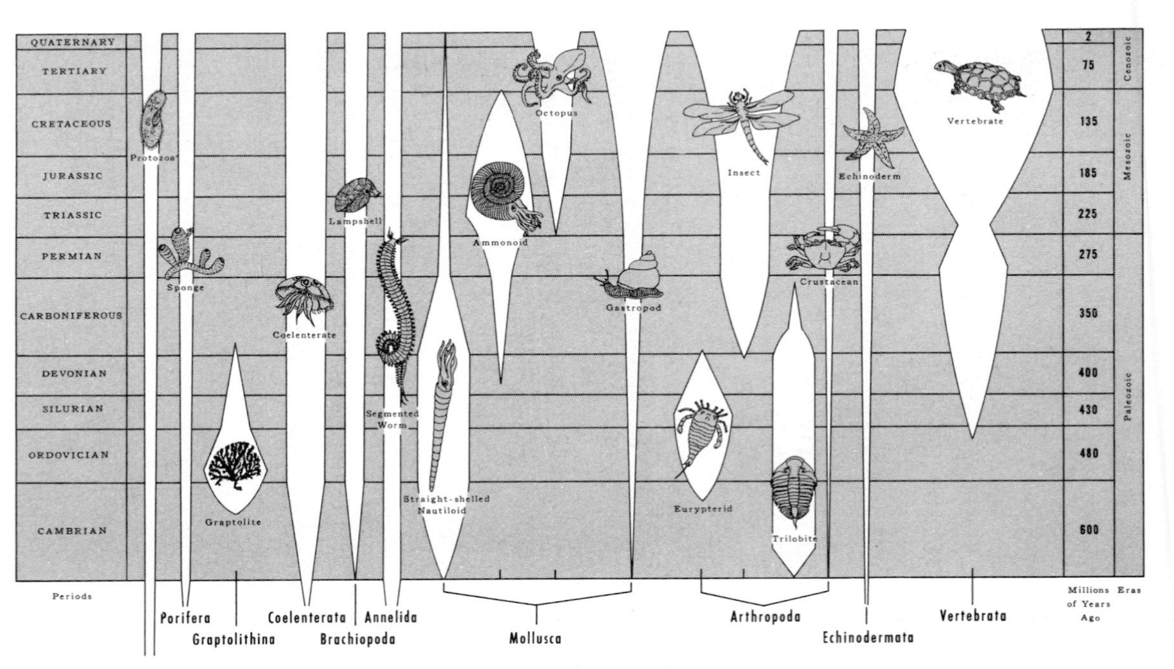

Figure 14.6 Historical record of animal life. Many of the invertebrate groups were already highly diversified and abundant in the Cambrian, the first period of the Paleozoic era, approximately 600 million years ago. The Paleozoic era is often called the *Age of Invertebrates,* with its multitude of nautiloids, eurypterids, and trilobites. Brachiopods with hinged valves were the commonest shellfish of the Paleozoic seas. In the Mesozoic era, air-breathing insects and vertebrates, notably the widely distributed reptiles, held the center of the stage. The Mesozoic seas were populated with large, shelled ammonoids, now extinct. Warm-blooded vertebrates (birds and mammals) became prominent in the Cenozoic era, and man himself arrived on the scene in the closing stages of this era (see chapter 15).

Figure 14.7 Highly diversified assemblage of invertebrates of early Paleozoic seas. *Top:* Ordovician period. Large organism in foreground is a straight-shelled nautiloid. Other prominent forms are trilobites, massive corals, smaller nautiloids, and a snail. *Bottom:* Cambrian period. Conspicuous animals are the trilobite (center foreground), eurypterid (center background), and the jellyfish (left). Other animals include brachiopods, annelid worms, sea cucumber, and varied shelled forms. (*Top,* courtesy of Chicago Natural History Museum; *bottom,* courtesy of the American Museum of Natural History.)

each other's advantage. There is little doubt, for example, that the rise and spread of flowering plants fostered the diversification and dispersal of insects. As flowering plants became less and less dependent on wind for pollination, a great variety of insects emerged as specialists in transporting pollen.

The colors and fragrances of flowers evolved as adaptations to attract insects. Flowering plants also exerted a major influence on the evolution of birds and mammals. Birds, which feed on seeds, fruits, and buds, evolved rapidly in intimate association with the flowering plants. The emergence of herbivorous mammals coincided with the widespread distribution of nutritious grasses over the plains during the Cenozoic era. In turn, the herbivorous mammals furnished the setting for the evolution of carnivorous mammals. The interdependence between plants and animals continues to exist in nature today.

Evolutionary Stability

The multitude of different kinds of present-day organisms is impressive. Yet, the inhabitants of the world today are only a small percentage of the tremendous array of organisms of earlier periods. As we have seen, the fate of most lineages of organisms throughout time is extinction. Apparently, only those populations that can continue to adapt to changing environmental conditions avoid extinction. Yet, some types of organisms have not changed appreciably in untold millions of years. Long-standing stability of organization seems antithetical to the concept of evolution. The opossum has survived almost unchanged since late Cretaceous, some 75 or more million years ago. The horseshoe crab, *Limulus,* is not very different from fossils uncovered some 500 million years ago. The maidenhair, or ginkgo, tree of the Chinese temple gardens differs little from its ancestors 200 million years back. The treasured ginkgo has probably existed on earth longer than any other present tree. Darwin called the ginkgo "a living fossil." We have no adequate explanation for such unexpected stability of organization. Perhaps some organisms have reached an almost perfect adjustment to a relatively unchanging environment. One thing, however, is certain: such stable forms are not at all dominant in our present-day world. One of the dominant forms today is man, a mammal that has evolved rapidly in a relatively short span of years.

Selected Readings

Adler, I. 1957. *How life began.* New York: The New American Library.

Barghoorn, E. S. 1971. The oldest fossils. *Scientific American,* May, pp. 30–42.

Calvin, M. 1969. *Chemical evolution.* New York: Oxford University Press.

Cohen, S. S. 1970. Are/were mitochondria and chloroplasts microorganisms? *American Scientist* 58:281–89.

Cohen, S. S. 1973. Mitochondria and chloroplasts revisited. *American Scientist* 61:437–46.

Eglinton, G., and Calvin, M. 1967. Chemical fossils. *Scientific American,* January, pp. 32–43.

Fingerman, M. 1975. *Animal diversity.* New York: Holt, Rinehart and Winston.

Fox, S. W., and Dose, K. 1972. *Molecular evolution and the origin of life.* San Francisco: W. H. Freeman and Co.

Glaessner, M. F. 1961. Precambrian animals. *Scientific American,* March, pp. 72–78.

Hanson, E. 1964. *Animal diversity.* Englewood Cliffs, N.J.: Prentice-Hall.

Jukes, T. H. 1968. *Molecules and evolution.* New York: Columbia University Press.

Keosian, J. 1967. *The origin of life.* New York: Reinhold Publishing Corp.

Margulis, L. 1970. *Origin of eucaryotic cells.* New Haven: Yale University Press.

Miller, S. L. 1953. A production of amino acids under possible primitive earth conditions. *Science* 117:528–29.

Oparin, A. I. 1953. *The origin of life.* New York: Dover Publications.

Orgel, L. E. 1973. *The origins of life.* New York: John C. Wiley & Sons.

Ponnamperuma, C. 1972. *The origins of life.* New York: E. P. Dutton & Co.

Romer, A. S. 1945. *Vertebrate paleontology.* Chicago: University of Chicago Press.

Rush, J. H. 1963. *The dawn of life.* New York: The New American Library, Signet Library.

Smith, K. M. 1962. *Viruses.* Cambridge: Cambridge University Press.

Stirton, R. A. 1959. *Time, life, and man.* New York: John C. Wiley & Sons.

Wald, G. 1954. The origin of life. *Scientific American,* August, pp. 44–53.

Walton, J. 1953. *An introduction to the study of fossil plants.* London: A. & C. Black.

15

Emergence of Man

Man has unique attributes; but he is nonetheless an animal and the product of the same natural evolutionary forces that have shaped all animal life. There is almost universal unanimity that the closest relatives of man are the apes. The line leading ultimately to man diverged from the ape branch during Tertiary times. There are several candidates among Tertiary fossil, manlike apes that qualify for the position ancestral to man. One often hears stated, hesitatingly and perhaps in the form of an apology, that man is not really closely related to the apes but merely shares a very distant ancestry with the apes. By indirection, the evasive idea is conveyed that our remote generalized ancestor would not be at all apelike. This is sheer deception. There is simply no blinking the question that modern man's ancestors and modern ape's ancestors were close relatives, and not distant relatives.

Primate Radiation

Primates, the order to which man belongs, underwent adaptive radiation when it first arose in Cenozoic times, approximately 70 million years ago. The primates are primarily tree dwellers; only man is fully adapted for life on the ground. Many of the noteworthy characteristics of the primates evolved as specializations for an arboreal mode of life. Depth perception is important to a tree-living animal; the majority of the primates are unique in possessing binocular, or stereoscopic, vision, wherein the visual fields of the two eyes overlap. The hands evolved as prehensile organs for grasping objects, an adaptation later useful to man for manipulating tools. The use of the hands to bring objects to the nose (for smelling) and to the mouth (for tasting) was associated with a reduction of the snout, or muzzle, and a reduction of the olfactory area of the brain. Primates generally have a poor sense of smell. Closely associated with enhanced visual acuity and increased dexterity of the hands was the marked expansion of the brain, particularly the visual and memory areas of the brain. Progressive enlargement of the brain culminated, in man, in the development of higher mental faculties.

The primate stock arose and differentiated from small, chisel-toothed, insectivorous ancestors (fig. 15.1). The Asiatic tree shrew, *Tupaia,* an agile tree climber that feeds on insects and fruits in tropical forests, survives as a model for the ancestor of the primates. Most authorities place the tree shrew in the mammalian order Insectivora, while a few hold the view that the tree shrew is a primitive, generalized member of the primates. The order Primates is generally divided into two groups or suborders, the *Prosimii* and the *Anthropoidea.* The prosimians are arboreal and largely nocturnal predators; they include such tropical forms as the lemurs, the lorises, and the tarsiers. As adaptations to an arboreal-nocturnal niche, these prosimians evolved

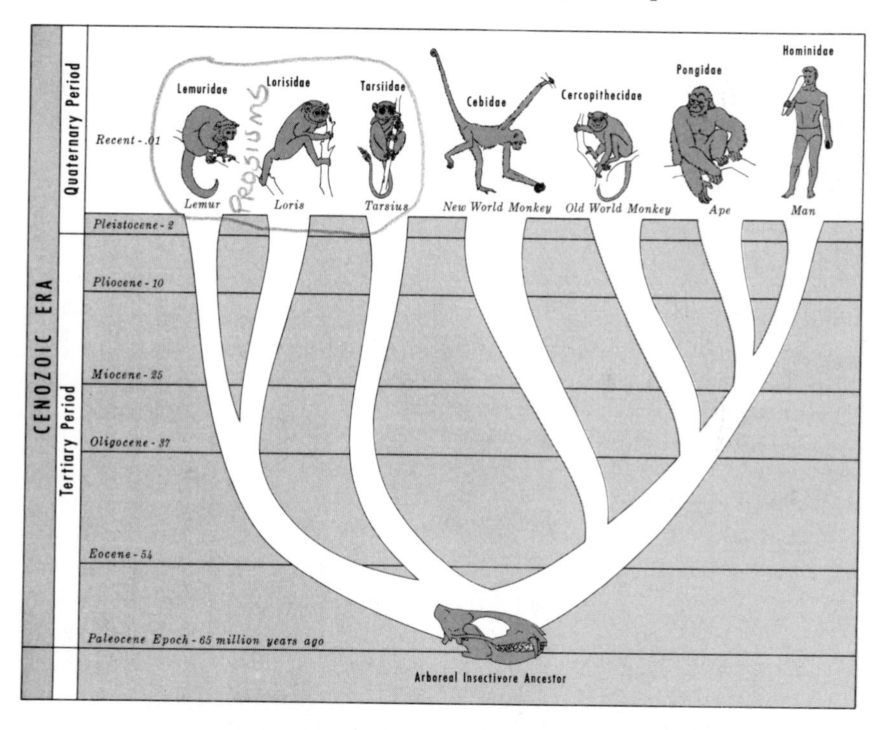

Figure 15.1 Adaptive radiation of the primates from a basic stock of small, insect-eating placental mammals, the Insectivora (whose living kin include the shrews, moles, and hedgehogs). Based on paleontological and anatomical evidence, the pongid (ape) and hominid (man) lineages diverged during the Miocene epoch, about 20 million years ago. Estimates based on biochemical data (e.g., comparative immunology of serum albumin) are presently at variance with the fossil data. The biochemical evidence suggests that the lineage leading to man split off from the ape line only about 4 million years ago.

large, forwardly placed eyes and strong, grasping hands and feet. The lemurs are largely confined to Madagascar, the lorises are found principally in eastern Asia, and the tarsiers are limited to southeastern Asia. The tarsiers represent the most advanced group of prosimians, and may be said to foreshadow the higher primate trends of the Anthropoidea.

The more advanced primates, the anthropoids, are composed of the New World monkeys, the Old World monkeys, the apes, and man. All are able to sit in an upright position, and thus the hands are free to investigate and manipulate the environment. The monkeys are normally quadrupedal, running along branches of trees on all fours. Among the apes, gibbons habitually use their arms for hand-over-hand swinging in a motion known as brachiation. The great apes (orangutan, chimpanzee, and gorilla) can maintain prolonged semierect postures. When on the ground, the chimpanzee and gorilla are "knuckle-walkers," the weight being placed on the knuckles of the hands rather than on the extended fingers. Man alone is specialized for erect bipedal locomotion.

The anatomical and physiological features that distinguish modern man from the living great apes are comparatively few. The resemblances in skeletal structures, muscular anatomy, physiological processes, serological reactions, and chromosome patterns are all strikingly close. The close relation of man and the apes shows up clearly in molecular comparisons of the alpha chain of the hemoglobin molecule. In the sequence of 146 amino acids of the alpha chain, there is only one difference in the composition of the amino acid residues between man and the gorilla. This contrasts sharply with differences, for example, of 19 amino acid residues in the alpha chain between man and the pig, or differences of 26 amino acid residues between man and the rabbit.

The pronounced differences between man and the apes relate mainly to locomotory habits and brain growth. Man has a fully upright posture and gait, and an enlarged brain. The cranial capacity of a modern ape rarely exceeds 600 cubic centimeters, while the average human cranial capacity is 1,350 cubic centimeters. Much of man's mastery of varied environments has been the result of his superior intelligence, gradually acquired throughout evolution.

When we speak of man, we inevitably think of him as he exists today. Present-day man is certainly different from his predecessors, in much the same manner that the modern horse is different from his forerunners. Thus, when a Pleistocene fossil specimen is designated an "ape-man" or a "near-man," it should not be imagined that such an extinct form possessed the qualities of man as we know him today. It is important to recognize that there have been different kinds of men. The evolutionary process of adaptive radiation led to a family of men, recognized formally as the Hominidae, of which modern man—*Homo sapiens,* or "man the wise"—is only one member and the sole survivor.

Fossil Preman

Fragmentary remains have been uncovered of apelike primates that inhabited the Old World during Miocene and early Pliocene times, spanning roughly the period from 25 million to 10 million years ago (fig. 15.2). Most are

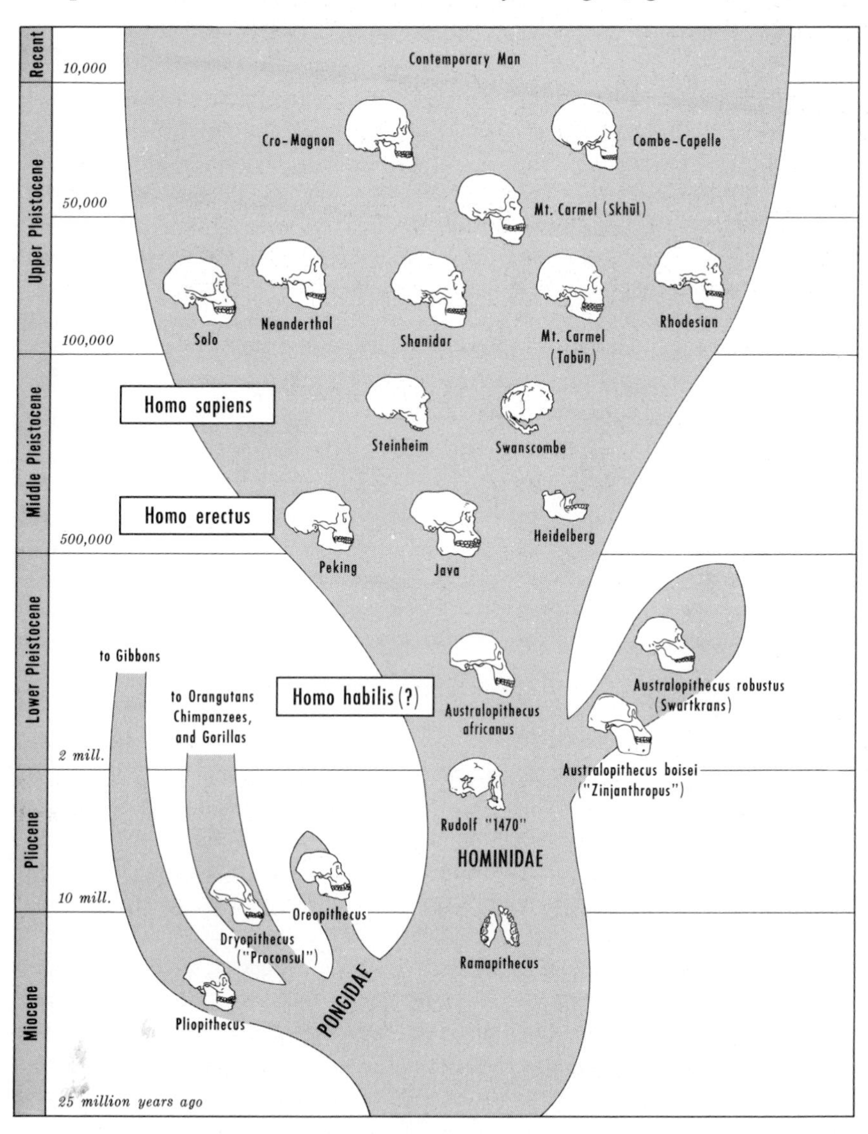

Figure 15.2 Geologic past of man. The late Miocene epoch witnessed the separation of the pongid (ape) and hominid (man) assemblages. The evolution of man was not limited to a single lineage. Several branches forked off from the main stem and were not directly ancestral to modern man, *Homo sapiens*.

clearly members of the ape family (Pongidae), such as *Pliopithecus,* a Miocene gibbonlike creature that is generally regarded as ancestral to today's gibbons. A fossil form that once aroused lively debate is *Oreopithecus,* unearthed in Italian coal beds that date back to the early Pliocene, approximately 10 million years ago. In the 1950s, the Swiss scholar Johannes Hurzler championed the view that *Oreopithecus* is a primitive hominid in the direct line of man's ancestry. The current consensus holds that *Oreopithecus* is an aberrant offshoot of the pongid stock, an evolutionary dead end (fig. 15.2).

Miocene sediments of Europe, Africa, and Asia (especially of India) have yielded remains of *Dryopithecus,* the oak-ape, so called because of the presence of oak leaves in the fossil deposits. Known collectively as the dryopithecines, these primitive oak-apes were the early forerunners of the modern orangutan, chimpanzee, and gorilla. The African variety, *Dryopithecus africanus*—first named *Proconsul*, after a famous chimpanzee at the London Zoo—may not be far removed from the common stock from which apes and men arose (fig. 15.2). Although primarily a tree dweller, *Dryopithecus africanus* apparently wandered on the ground and may have availed himself of vegetable foods in the open grasslands (fig. 15.3). The transition from tree dwelling to ground living might well have first appeared at this time. In the Miocene epoch, the great expanse of tropical rain forest in eastern Africa dwindled, leaving patches of dry wooded areas separated by open bushy grasslands. The capacity for ground walking may have evolved as an adaptation enabling arboreal forms to cross the open plains in passing from one patch of woodland to another. Thus, *Dryopithecus africanus* may have become a ground dweller in order to continue to live successfully in the trees. This is a seeming paradox, parallel to the intriguing conjecture that water-living vertebrates evolved terrestrial habits so that they could retain, and not abandon, their aquatic mode of life.

An impressive series of Miocene fossils has been placed in the genus *Ramapithecus*. An upper jaw, with unmistakable hominid dentition, was first found by G. Edward Lewis in 1935 in the Silawik Hills of northern India. This fragmentary specimen was named after the Indian god, Rama. In the 1960s, Louis S. B. Leakey uncovered jaw fragments of *Ramapithecus* in eastern Africa (Kenya) in deposits of a geological age of 14 million years ago. These important fossil jaws show reduced canine teeth, relatively small incisors, and hominidlike low-crowned molars. *Ramapithecus* evidently ranged widely throughout Africa and Asia in late Miocene and early Pliocene times. This fossil specimen may be close to the base of the stem of the hominid family. That is, this fossil form may mark the point where the hominid lineage separated from the pongid assemblage (fig. 15.2). The roots of man's family tree apparently date back 20 to 25 million years.

Several authors accept *Ramapithecus* as the earliest member of the family of men (Hominidae). The hominid status implies that this ancestral

Figure 15.3 Reconstruction of *Dryopithecus africanus,* an apelike type that prowled East Africa 20 to 25 million years ago. This pongid apparently led an agile life both on and off the ground. [Painting by Maurice Wilson; courtesy of the British Museum (Natural History).]

form was bipedal and ventured into open country to forage for food. There is no evidence that *Ramapithecus* was a hunter or tool user, or that he was a habitual ground-dwelling biped. It would seem that the distinctive dental adaptations of hominids evolved before the bipedal mode of locomotion became habitual or methodical. The evolutionary development of the capacity for efficient bipedalism probably required millions of years.

Ape-Men or Near-Men

In 1924, an epochal discovery in South Africa was announced by Raymond A. Dart, an Australian anatomist at the University of Witwatersrand in Johannesburg. In a Pleistocene limestone quarry near the small village of Taung in Bechuanaland was found the fossilized skull of a juvenile, corresponding to the skull of a child of about six years. The little Taung skull bears some resemblance to the skull of a young chimpanzee, but many of its components, notably the teeth, show pronounced affinities to man. Dart named the remarkable skull *Australopithecus africanus* (*austral,* for "south," and *pithekos,* for "ape"). Dart asserted that the southern ape-man was related to the ancestral stock of man rather than to the great apes (fig. 15.2). In other words, the Taung fossil represented an early member of the Hominidae, the family of men, rather than of the Pongidae, the family of apes.

Dart's declaration, which several scientists initially vigorously disputed, was fortified by findings in the 1930s by the late Robert Broom, a Scottish paleontologist. Adult skulls of *Australopithecus* were dug out from caves in Sterkfontein, Kromdraai, and Swartkrans in South Africa. The adult skulls confirmed the hominid anatomical pattern seen in the juvenile Taung cranium. The several new fossil forms were originally given different names. However, in recent years, it has become customary to refer to the South African ape-man collectively as the australopithecines. They were short, four to five feet in height, with a small ape-sized brain (cranial capacity range of 450 to 600 cubic centimeters). Nonetheless, the australopithecines stood upright, walked bipedally, and dwelt in open country (fig. 15.4). These circumstances nullify the popular view that man was an intelligent animal when he first came down out of the trees. It seems clear that erect bipedal locomotion on the ground preceded the development of a large complex brain.

The australopithecines are decidedly early representatives of the hominid lineage. However, spirited debate exists over which of the australopithecines occupy a prominent place in the direct ancestry of man. Current evidence favors the idea that there were two distinct species of australopithecines—the light-jawed, lightly built *Australopithecus africanus* and the heavy-jawed, larger *Australopithecus robustus* (fig. 15.2). The former species was not over four feet tall and weighed no more than 60 pounds, whereas the latter species was a foot taller and at least 30 pounds heavier. When two contemporaneous species occupy the same habitat, ecologists have observed that the potential competitors become differentially specialized to exploit different components of the local environment. Thus, direct competition for food resources is minimized and the two species are able to coexist. It is thought that *Australopithecus robustus* became or remained exclusively vegetarian, whereas *Australopithecus africanus* increasingly supplemented its diet with animal food. The dietary difference is supported by the finding that *A. africanus* had smaller molars than *A. robustus.*

If, as it appears likely, the two *Australopithecus* species did coexist at the same time in the same region, then only one of the two could have been the progenitor of a more modern species of man. It has been suggested that the vegetarian *A. robustus* perished without leaving any descendants, and that

Figure 15.4 Reconstruction of *Australopithecus,* an ape-man of about 1 million years who stood upright, walked bipedally, and dwelt in open country. (Painting by Maurice Wilson; courtesy of the British Museum [Natural History].)

A. africanus was the forebear of a more advanced hominid. The dietary change to carnivorism may have represented one of the most important steps in transforming a bipedal ape into a tool-making and tool-using man.

The South African finds have been supplemented and extended by fossil discoveries in eastern Africa since 1959 by Louis and Mary Leakey at the 25-mile-long Olduvai Gorge in Tanzania, and in more recent years by Richard Leakey at the eastern shore of Kenya's Lake Rudolf. At Olduvai Gorge, the Leakeys (Louis and Mary) in 1959 uncovered bony fragments of a robust australopithecine, characterized by extremely massive jaws. This heavy-jawed fossil form was called *Zinjanthropus,* or the Nutcracker Man. Fossil remains of *Zinjanthropus* were found in strata judged, by a new potassium-argon dating method (instead of the conventional uranium-lead technique), to be about 1.8 million years old. Most scientists today agree that *Zinjanth-*

ropus is essentially an eastern African variety of *Australopithecus robustus*. Others contend that *Zinjanthropus* has more exaggerated, or coarser, features than *A. robustus,* and warrants recognition as a separate species, *Australopithecus boisei.* It would thus represent a third species in the australopithecine complex.

In the 1960s, Louis and Mary Leakey found remains of a light-jawed hominid that they claimed was more advanced than *Australopithecus africanus.* Estimates of the brain capacities of the skulls averaged 637 cubic centimeters. This light-jawed type of Olduvai Gorge was said to be the first civilized or humanized man, deserving of the rank of *Homo,* namely *Homo habilis* (fig. 15.2). The specific name *habilis* means "able, handy, mentally skillful, vigorous," from the inferred ability of this man to make stone tools. The recognition of *Homo habilis* indicates that this primitive human being was evolving alongside the less hominized australopithecines, and lived side-by-side with them. The coexistence of *Homo habilis* and the australopithecines has been the subject of much concern among evolutionists. Some writers argue that *Homo habilis* did not cross the threshold between the prehuman and human grades, and represented only an advanced member of *Australopithecus africanus.*

In 1972, Richard Leakey unearthed from the dessicated fossil beds of Lake Rudolf a nearly complete skull that has been placed at 2.8 million years. The specimen has been cautiously designated only by its museum identification number—"1470." The cranial capacity of the "1470" skull measures 780 cubic centimeters, which is significantly larger than any australopithecine specimen. Skull "1470" clearly belongs to the genus *Homo,* although no species name has yet been designated. It may represent *Homo habilis,* and it may be the linear ancestor of *Homo erectus.*

In 1975, Mary Leakey and her co-workers discovered fossil jaws and teeth in sediments at a site 25 miles south of Olduvai Gorge called Laetolil. These hominid jaw fragments bear a strong resemblance to the mandible and teeth of skull "1470." The Laetolil specimens are impressive in that they have been dated at 3.35 to 3.75 million years. Mary Leakey has classified the jaw fragments as the remains of the eastern African *Homo habilis.* If this is the case, the existence of *Homo habilis* has been extended back to at least 3.5 million years ago; accordingly, this human species is older than any known australopithecine. This would indicate that *Homo habilis* is not the direct descendant of the australopithecines. It may be, however, that an older, as yet undiscovered *Australopithecus* (from Pliocene sediments) gave rise to *Homo habilis.*

Evolution of Hunting

The lower Pleistocene hominids (*Australopithecus africanus* or *Homo habilis,* depending on one's views) fabricated crude tools out of pebbles. The ability to fashion and wield tools followed, or accompanied, the emancipation of the hands that bipedal locomotion made possible. The open African savanna offered to the early plains-dwelling hominids a new food resource; namely, herbivorous animals of all sizes adapted to the grassland habitat. A meat-eating hominid, however, cannot tear the thick hide of an antelope or a deer with his fingernails and teeth. A strong selective premium was placed on the acquisition of new manipulative skills for more efficient tool or weapon construction. To shape a stone into a sharp-pointed blade or to fashion a stick into a spear depends on improved mental capacities. Increased and more efficient tool use brought about selective pressures favoring increased brain size and complexity. Although these views lean heavily on speculation, it appears likely that adoption of the hunting habit shaped the behavior pattern of the human species. Hunting became more than a mere technique; it became a way of life. As discussed in a subsequent section, the hunting habit led to a division of labor between the sexes, the introduction of food sharing, and the eventual establishment of a unique family organization.

Early True Man

The famous Java man, first described as *Pithecanthropus erectus* ("erect ape-man"), undoubtedly had crossed the threshold between prehuman and human grades (fig. 15.2). This primitive human being was discovered at Trinil, Java, in 1894 by Eugene Dubois, a young Dutch army surgeon. Dubois had been profoundly influenced by the writings of Charles Darwin, and had become imbued with the idea that he could find the origins of man in the Far East. He surprised the world with the discovery of this early human. Curiously, Dubois, in his later years, inexplicably doubted his own finding and contended that *Pithecanthropus erectus* was merely a giant manlike ape. In the 1930s, additional fossil finds of *Pithecanthropus* were unearthed in central Java by the Dutch geologist G. H. R. von Koenigswald. The new findings confirmed the human status of the pithecanthropines.

The pithecanthropines lived during middle Pleistocene times, between 500,000 and 300,000 years ago (fig. 15.2). These low-browed men were toolmakers and hunters who had learned to use fire. They probably had some powers of speech. Their ability to exploit the environment is reflected in the expanded size of the brain. The cranial capacity of the pithecanthropines was in the range of 775 to 1,000 cubic centimeters.

In the 1920s, elaborate excavations undertaken by the Canadian anatomist Davidson Black of caves in the limestone hills near Peking, China, led to the discovery of another primitive man, *Sinanthropus pekinensis,* or Peking man (fig. 15.5). The cranial capacity in the sinanthropines varied from 900 to 1,200 cubic centimeters. They fashioned tools and weapons of stone and bone, and kindled fire. There is a strong suspicion that *Sinanthropus* was cannibalistic and savored human brains, for many of the fossil braincases show signs of having been cracked open from below.

Figure 15.5 Peking man left his remains about 400,000 years ago in limestone caves in northern China. He kindled fire, and fashioned tools of stone and bone. (Painting by Maurice Wilson; courtesy of the British Museum [Natural History].)

Java man and Peking man were originally each christened with a distinctive Latin name, *Pithecanthropus erectus* and *Sinanthropus pekinensis,* respectively. There is, however, no justification for recognizing more than the single genus of humans, *Homo.* Accordingly, modern taxonomists have properly assigned both Java man and Peking man to the genus *Homo.* Moreover, the morphological differences between these two fossil men are readily within the range of variation that we observe in living populations today. These forms thus represent two closely related geographic races (subspecies) of the same species. Lastly, both Java and Peking men are distinct enough from modern man (*Homo sapiens*) to warrant being placed together in a different species, *Homo erectus.*

These nomenclatural changes may appear to be trivial, but the implications are great. One important implication is that human populations at any one time level were differentiated into geographical races of one species—not into distinct species or even genera. We can envision, for example, that approximately 500,000 years ago, there existed a single widespread species of man, with eastern populations represented by the Java and Peking variants (*Homo erectus erectus* and *Homo erectus pekinensis,* respectively) and the western populations constituted by types resembling Heidelberg man (found in an early Pleistocene sand deposit near Heidelberg, Germany). The suspected wide distribution of *Homo erectus* has been confirmed by the recent discoveries of this early type of man in northern Africa (Ternifine, Algeria) and eastern Africa (Olduvai Gorge, Tanzania). Evidently then, populations of *Homo erectus* spread successfully through the continents, from the tropical regions of Africa to southeast Asia.

Evolution of Human Society

The hunting-gathering economy of *Homo erectus* necessitated or permitted collaborative, or cooperative, interactions of individuals. Adult males became increasingly interdependent in their hunting efforts. *Homo erectus* hunted in small groups in open country, using wooden spears, clubs, and stones. Dismembering and sharing the animal prey or kill became a fundamental disposition of the group or band. The sharing of food regularly is a social achievement unique to man; only rarely do apes share food. To this day, humans express sentiments of comradeship and trust by coming together in common meals.

There is no evidence that the early hominid female participated in the hunting of large game; the adult female was increasingly encumbered with a fetus or by the care of the young, or by both. The long period of dependency of the young strengthened mother-child bonds but also restricted the mobility and activity of the woman. It appears likely the female remained at home as a food-gatherer and domestic while the male alone engaged in hunting. The immobility of the female and the prolonged immaturity of the young, coupled with the limitations imposed on the male in the number of females he could possibly support, radically transformed a basically polygamous society into a monogamous structure.

A primary human innovation was thus relatively permanent pair-bonding, or monogamy. Sustained pair-bonding proved to be advantageous in several respects. It served to reduce sexual competition and the prolonged sexual contact increased the probability of leaving descendants. The heterosexual pair-bonding relationship became fortified as the estrus cycle of the female became modified into a condition of continuous sexual receptivity. The sustained sex interests of the partners made possible by the obliteration of estrus in the female increased the stability of the family unit, and facilitated the development of permanent family-sized shelters for rest, protection, and play. In essence, strong interpersonal bonds between a male, a female, and their children became the basis of the uniquely human family organization. The human male had become incorporated into the mother-young social group of the monkeys and apes.

Human language was fostered as the hunters recorded experiences with each other and transmitted information to their mates and children. In turn, the women conveyed events on the domestic scene to the returned hunters. Speech favored cooperation between local groups, and the fusion of small groups into larger communities. Speech also fostered the successful occupation of one geographical area after another. The exchange of ideas over wide areas permitted human cultures of great complexity to develop. Evidently then, early man's assumption of the hunting habit was preeminent in shaping the social organization of early (and later) human society.

Emergence of Modern Man

The classic Neanderthal man was first unearthed in 1856 in a limestone cave in the Neander ravine near Dusseldorf, Germany (fig. 15.6). It is one of the best known of fossil men, having been subsequently found at numerous widely separate sites in Europe, particularly in France. Neanderthal man was a cave dweller, short (about 5 feet) but powerfully built, with prominent facial brow ridges, and a large brain with an average capacity of 1,450 cubic centimeters (as opposed to 1,350 cubic centimeters in modern man). Neanderthal man first arose some 100,000 years ago (fig. 15.2). He roamed over Europe up to about 40,000 years ago (Upper Pleistocene), and then he dramatically disappeared. He was replaced by men of a modern type, much like ourselves, which have been grouped under the common name of Cro-Magnon.

Figure 15.6 Neanderthal man, a rugged cave-dweller who roamed Europe and the Middle East about 75,000 years ago. (Painted by Maurice Wilson; courtesy of the British Museum [Natural History].)

The transition from Neanderthal man to Cro-Magnon man is problematical. Prior to the time of the Neanderthalers themselves, about 200,000 to 100,000 years ago, there emerged types of men, such as Swanscombe from England (fig. 15.7) and Steinheim from Germany that did not conform to the classic Neanderthal type. Indeed, the Swanscombe and Steinheim skulls (fig. 15.2) are not markedly different from the skull of modern man. These Middle Pleistocene skulls from Europe clearly belong to *Homo sapiens* rather than *Homo erectus*. Is it possible that a modern type of man arose before Neanderthal? We have grown accustomed to the idea that Neanderthal man was our direct ancestor, but now it appears that we may be closer to the truth by considering him a sterile offshoot. The picture is far from clear. In the Middle East about 100,000 to 30,000 years ago there existed an exceptionally heterogeneous group of men. These are represented by the Palestinian Mount Carmel man, dug out of caves at Tabūn and Skhūl, and the Shanidar man, excavated from caves in the mountains of northern Iraq (fig. 15.2). These men ranged from individuals with almost typically western European Neanderthal features to individuals that are barely distinguishable from modern man.

Another element of uncertainty is the relation of the European Neander-

Figure 15.7 Reconstruction of the life of Swanscombe man from England, one of the earliest members of *Homo sapiens*. Swanscombe man spread across the plains of northern Europe about 200,000 to 100,000 years ago. (Painting by Maurice Wilson; by permission of the British Museum [Natural History].)

thal man to other Upper Pleistocene men in widely scattered parts of the world—Rhodesian man of central Africa and Solo man of central Java (fig. 15.2). Some authorities have considered Rhodesian and Solo men to be the geographical equivalents of Neanderthal man of Europe. Solo man has been characterized as a "tropical Neanderthaler." This would lead us to believe that the classic Neanderthal man was widely distributed in the Old World. These findings present knotty problems.

Status of Neanderthal Man

Authorities do not all agree on the status of Neanderthal man. The populations of Neanderthal man in Europe disappeared rather abruptly and were replaced by a modern group (Cro-Magnon) that definitely belongs to *Homo sapiens*. Does this indicate that two separate species, *Homo neanderthalensis* and *Homo sapiens,* actually existed together in Europe, and that the latter species displaced the former without hybridization occurring between the two? The evolutionist Theodosius Dobzhansky has expressed perceptive views on this question. Dobzhansky calls attention to the Mount Carmel fossil populations in Palestine, particularly the one found at Skhūl. This extraordinary fossil assemblage consists of individuals ranging from classical European Neanderthal types to forms closely resembling *Homo sapiens.* This suggests a racial mixture of the two groups. It appears, then, that the European Neanderthal man and the Middle East modern men were not reproductively incompatible, as would be two species, but rather were races of the same species (*Homo sapiens*). In other words, the Mount Carmel locality represented a zone of intergradation of a kind usually found at the boundaries of geographic races. The emergence of Cro-Magnon man may have resulted, at least to some extent, from the amalgamation of the European Neanderthal race and the Middle East modern-type race invading Europe.

Although this interpretation seems reasonable, some authorities assert that the classic Neanderthalers were not directly involved in the ancestry of modern man. Clark Howell, professor of anthropology at the University of Chicago, suggests that little, if any, opportunity existed for the exchange of genes between the Neanderthal population of western Europe and populations in the Middle East or elsewhere. During the Pleistocene epoch, glacial ice sheets covered many parts of Europe. It may be that the Neanderthalers of western Europe were geographically isolated by the Ice Age of 100,000 years ago. Neanderthal man may have perished in isolation before modern man arrived, or may have been overrun by more progressive Middle East contemporaries who had spread into western Europe. Under this viewpoint, Neanderthal man represented an evolutionary dead end in man's ancestry. Modern man evolved independently of Neanderthal man, and arose by separate origin via the middle Pleistocene *sapiens*like stock—Steinheim man and Swanscombe man—and the more advanced Mount Carmel men in Upper Pleistocene.

Cro-Magnon man, a representative of our own species, *Homo sapiens,* can be traced back about 35,000 years. His remains have been found in many sites in western and central Europe. The Cro-Magnon men showed individual differences, just as man today exhibits individual variation. One of the notable variants is Combe-Capelle man from France (fig. 15.8). Little is known of modern man in other continents during the time that Cro-Magnon flourished in Europe. Modern man may have been cradled in Asia or Africa.

Pleistocene Overkill

Cro-Magnon man was an accomplished hunter, so much so that he has been held responsible for the high rate of destruction and extinction of the mammalian fauna—such as the giant sloth, mammoths, saber-toothed cat, and giant ox—in the upper Pleistocene. The large-scale annihilation of many game mammals by man has been called *Pleistocene overkill* by the

Figure 15.8 Cro-Magnons, representatives of our own species, *Homo sapiens,* can be traced back about 35,000 years ago in Europe. (Painting by Maurice Wilson; courtesy of the British Museum [Natural History].)

evolutionist Paul S. Martin. Martin has noted that more than a third of the genera of large mammals met extinction in Africa about 50,000 years ago, and that nearly all of the larger game genera in North America perished 12,000 to 15,000 years ago. The latter event coincided with the migrations of early Mongoloid bands into North America via the Bering Strait. So devastating were the hunting activities of man in North America that only a few forms of larger mammals survived, notably the bison and pronghorn antelopes. Such drastic reductions in game animals throughout the world may have been one of the precipitating factors in man's transition some 10,000 years ago from a hunting economy to an agricultural economy.

Origin of Races of Man

We do not know the birthplace of modern man. Nevertheless, we do *not* believe that Cro-Magnon man originated simultaneously in widely different parts of the world. Rather he arose in one place, then migrated to various regions of the globe and became differentiated into geographical races. This is the orthodox and established pattern of geographical origin of races. There is, however, a school of thought, chiefly identified with Franz Weidenreich and Carleton Coon, that conceives of the modern races of man as descending from different ancient hominid lineages evolving independently of one another. Thus, as seen in figure 15.9, Java man was the early progenitor of the Australoid race; Peking man was forerunner of the Mongoloid race; Rhodesian man gave rise ultimately to the Negroid race; and a Middle East type, perhaps Mount Carmel, led to the Caucasoid race. According to this theory, the races of man would need to be considered older than the species *Homo sapiens* itself.

At this point, we may digress to consider the phenomenon of parallel evolution, or *parallelism,* whereby two organisms acquire similar characteristics independently of one another although they have stemmed from related ancestral stocks. An illustrative case would be the structural resemblances of New World and Old World porcupines, the familiar spine-bearing rodents. The New World porcupines are native to South America; their counterparts in North America are relatively recent immigrants from the South American forests. The Old World porcupines are common in Africa and have spread into southern parts of Eurasia. Some students of the subject believe that the South American porcupines are direct descendants of the African forms. It has been suggested that the African porcupines crossed the South Atlantic on raftlike floating objects. A transoceanic dispersal route, however, seems improbable to several authorities. An alternative explanation, championed by the paleontologist Albert Wood of Amherst College, is that the South American and African porcupines have been derived independently of each other from an ancient generalized (nonporcupine) rodent stock that inhabited both the New World and the Old World. Thus, the South American

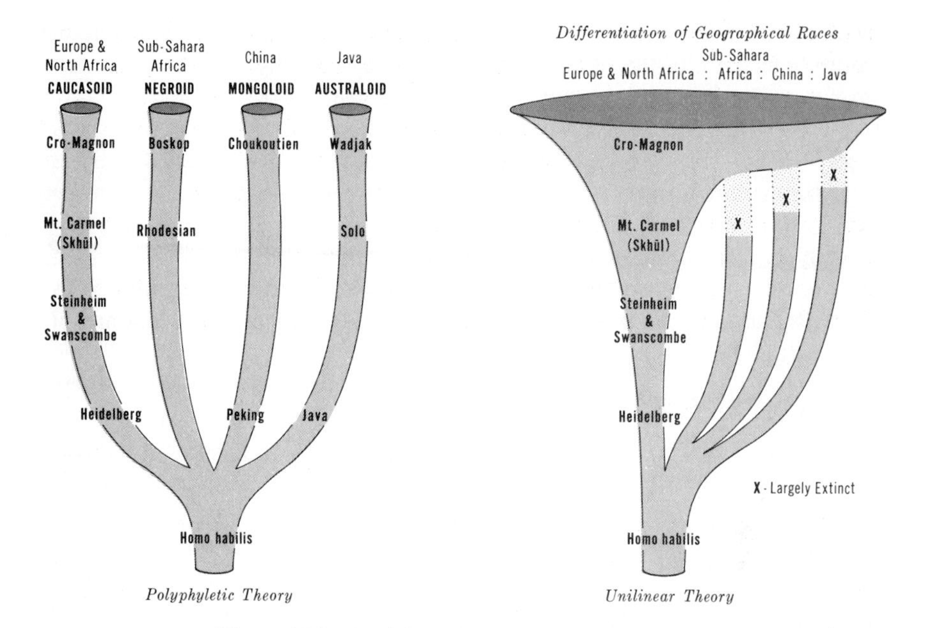

Figure 15.9 Racial origins of man. The polyphyletic hypothesis envisions a distant separation of the principal races of man. The four major geographical races had evolved independently and in parallel fashion over hundreds of thousands of years. This thesis that the major races of mankind can be traced far back in prehistoric antiquity is scarcely defensible in light of modern evolutionary concepts. The recent origin of the living races of man, as expressed by the unilinear theory, is more tenable. Racial differentiation occurred only after modern man (Cro-Magnon) arose and became distributed over a large area of the world. The unilinear theory does not exclude the persistence of some local traits from ancient regional lineages. Certain geographical populations reflect admixture of the immigrant Cro-Magnon group and the older indigenous stocks.

and African porcupines share a common ancestor (a primitive rodent stock), but independently, in parallel fashion, acquired structural resemblances.

Is it possible that widely separated geographical populations of early Pleistocene men followed independent, but parallel, courses of evolutionary development, much as the New World and Old World porcupines? Let us examine the situation carefully. By no stretch of the imagination could the South American and African porcupines be considered as *racial variants of one species*. These two porcupine groups are certainly reproductively incompatible. The Weidenreich-Coon School would argue that geographic separation of the different early hominid branches did *not* lead to reproductive isolation, as might be expected of long-standing populations that are spatially separated and that differentiate along independent lines. It is, however, exceedingly difficult to imagine how several hominid races, diverging in different parts of the world, can evolve independently and yet repeatedly in the same direction leading only to one species, *Homo sapiens*. The parallel origin of races is not hopelessly out of the question, but if it occurs, it must be the very rare exception to the normal process. Indeed, modern evolutionists are disposed to relegate the Weidenreich-Coon notion to parallel evolution of races to the category of the highly improbable.

There seems little doubt that *Homo sapiens* originated in a single area, then spread the world over and differentiated into regional populations. These regionally separated groups, once spatially separated to a large extent, have intermingled and intercrossed for untold thousands of years. The distinguishing features of the basic racial groups have become increasingly blurred by the countless migrations and intermixings. The whole world today is a single large neighborhood. Modern man potentially lives in one great reproductive community.

Problems of Present-Day Man

We have seen that the hunting way of life dominated man's existence for well over 600,000 years. Man discarded the arboreal, defensive, and herbivorous habits of his primate relatives and became a terrestrial, weapon-making, carnivorous predator. Man became endowed with genetic characteristics that enabled him to be successful as a predatory hunter. It is highly probable that many of the inherent adaptive characteristics of the hunting era have continued to persist in present-day man. Today, man lives for the most part in an environment of his own making. It is still an open question whether he can govern himself prudently in an environment that he has so drastically altered. Is it possible that contemporary man is unable to cope with himself or with his very recent environmental changes because those unique genetic propensities that were once highly adaptive in the hunting era are no longer adaptive in the modern era?

Selected Readings

Birdsell, J. B. 1974. *Human evolution*. Chicago: Rand McNally College Publishing Co.

Campbell, B. G. 1966. *Human evolution*. Chicago: Aldine Publishing Co.

Clark, W. E. LeGros. 1964. *The fossil evidence for human evolution*. Chicago: University of Chicago Press.

Eimerl, S., DeVore, I., and editors of Time-Life Books. 1965. *The primates*. New York: Time-Life Books.

Howell, F. C., and editors of Time-Life Books. 1965. *Early man.* New York: Time-Life Books.

Howells, W. W. 1959. *Mankind in the making.* Garden City, N.Y.: Doubleday & Co.

Howells, W. W. 1960. The distribution of man. *Scientific American,* September, pp. 112–27.

Howells, W. W. 1966. *Homo erectus. Scientific American,* November, pp. 46–53.

Lee, R. B., and DeVore, I. 1968. *Man the hunter.* Chicago: Aldine Publishing Co.

Medawar, P. B. 1961. *The future of man.* New York: The New American Library.

Montagu, A. 1962. *Man: his first million years.* New York: The New American Library.

Morris, D. 1968. *The naked ape.* New York: McGraw-Hill Book Co.

Morris, D. 1969. *The human zoo.* New York: Dell Publishing Co.

Napier, J. 1967. The antiquity of human walking. *Scientific American,* April, pp. 56–66.

Pearson, R. 1974. *Introduction to anthropology.* New York: Holt, Rinehart and Winston.

Pfeiffer, J. E. 1969. *The emergence of man.* New York: Harper & Row.

Pilbeam, D. R. 1972. *The ascent of man.* New York: Macmillan Co.

Shepard, P. 1973. *The tender carnivore and the sacred game.* New York: Charles Scribner's Sons.

Simons, E. L. 1964. The early relatives of man. *Scientific American,* July, pp. 50–62.

Washburn, S. L. 1960. Tools and human evolution. *Scientific American,* September, pp. 62–75.

Young, L. B., ed. 1970. *Evolution of man.* New York: Oxford University Press.

Index

(Page numbers in boldface refer to illustrations.)